ALL THAT GLITTERS

TRUTH BEHIND THE GLITTER

CONSPIRACY REVEALED

By Justin Time

Justin Time

This book is a work of non-fiction. While the events and facts presented in this book are true to the best of the author's knowledge, names and identifying details of individuals may have been changed to protect their privacy.

Copyright © 2024 Justin Time
ISBN: 9798338002506

-Imprint:
Independently published
All rights reserved

Cover design by: Grant Doubt Designs
Library of Congress Control Number:

Printed in the United States of America

Dedication

To the power of an idea, whose spark ignites the imagination and transforms the impossible into the inevitable.

As Einstein once said, "Imagination is more important than knowledge."

This book is dedicated to those who dare to dream, for it is in their visions that the future takes shape.

Dream On!

Inspirational Quotation

"To accomplish great things,

We Must;

Not only act,
But also dream;

Not only plan,
But also believe."

-Anatole France

Table of Contents

Foreword

- By an Anonymous Industry Insider

In a world where secrecy often fuels innovation, few could have imagined that something as small and seemingly inconsequential as glitter would carry such profound significance. My identity must remain hidden, as the delicate balance between progress and protection demands anonymity.

Yet, I cannot stay silent about the monumental impact that *All That Glitters* will have—not just on the industry, but on the world at large.

When Justin Time embarked on his investigation into the mysterious world of glitter, I observed from afar, bound by legal and corporate obligations that prohibited any direct or indirect involvement. I watched as he navigated a maze of misinformation, secrecy, and resistance, each step bringing him closer to uncovering a reality that few dared to acknowledge.

was that he might be discouraged 1 end—when the forces aligned :med too powerful to overcome.

But Justin's determination and tenacity proved stronger than the obstacles in his path.

Having had the opportunity to review a pre-published copy of this book, I can say with confidence that he has done a service not just to the industry, but to the world. By bringing this subject out of the shadows and into the light, he has ensured that what was once whispered in hushed tones will now be discussed openly, in living rooms, boardrooms, and legislative halls.

Make no mistake—this book is going to stir up a storm within the glitter industry. Initially, there will be outrage. Those who have guarded the secrets of this material for decades will be furious that certain truths have come to light, revealing what many believed should forever remain concealed. I expect intense discussions, confusion, and perhaps even a sense of betrayal among my peers.

However, these initial reactions will soon give way to a broader understanding. As the world begins to comprehend the revelations within these pages, it will become clear that Justin Time has done the industry a great service. What was once a niche product, confined to specific markets, will now emerge as a material of global importance.

The skills and expertise required to produce this highly specialized material will be in demand like never before.

Craftsmen and manufacturers who have labored in relative obscurity will find themselves thrust into the spotlight, their knowledge and capabilities sought after for new and innovative applications. Development contracts, research initiatives, and high-level collaborations will follow, transforming what was once considered an ordinary commodity into a crucial asset. The industry will not just survive this transition—it will thrive, creating new opportunities and redefining the role of glitter on the world stage.

This is not merely a shift in public perception; it's a reimagining of what glitter can be. No longer seen as a frivolous decoration or a micro-pollutant, glitter will now be recognized as a vital player in shaping global advancements across various sectors. Justin's work has ensured that what was once whispered in private conversations will now be openly discussed, debated, and embraced for its potential to influence the future in ways we could never have anticipated.

For those of us who have long understood the true potential of this material, this book is a catalyst for change. It is more than an exposé—it is a call to action that will elevate glitter from its humble beginnings to a new level of significance. The path ahead may be challenging, but it is one that offers immense opportunity for those willing to embrace this transformation.

In closing, I want to extend my deepest gratitude to Justin Time for his relentless pursuit of the truth. His dedication to uncovering the untold story of glitter has positioned it at the center of a worldwide conversation. Though I must remain in the shadows, I fully support this work and its message. Thanks to Justin's efforts, we are now on a path that promises to unlock possibilities we had only begun to imagine.

Written by a longtime industry insider and advocate of innovation, whose identity is and will remain anonymous due to legal restrictions advised, obtained, and secured by counsel.

Prologue

In the shadowy corners of the internet, where fact and fiction often blur into a perplexing haze, a whisper began to surface. At first, it was just that—a faint murmur, buried in obscure forums and dismissed in late-night conversations.

But as 1999 drew to a close, that whisper began to grow louder, gaining momentum, and capturing the imaginations of those who dared to listen. What was once a fleeting curiosity about the strange disappearance and stockpiling of glitter soon transformed into a full-blown phenomenon—one that would come to dominate online discussions and ignite a wildfire of speculation.

When I first stumbled upon the whisperings of a conspiracy involving glitter, I never imagined that it would lead me down a rabbit hole so deep that it would challenge my understanding of the world. What began as a curious fascination with an odd and seemingly insignificant mystery soon became an all-consuming quest for truth—a truth that, once revealed, has the potential to shake the very foundations of our reality.

The Glitter Conspiracy, as it has come to be known, isn't just another wild theory. It's a story with layers—layers that, as you'll discover, reveal far more than meets the eye. What started as an eccentric curiosity has evolved into a puzzle that challenges the very fabric of our understanding of power, secrecy, and influence.

Who is behind the massive, unexplained purchases of glitter? What could possibly justify such secrecy and subterfuge?

And, most chillingly, why has every attempt to uncover the truth been met with denial, deflection, and disinformation?

Throughout 2000 and well into 2024, this conspiracy spread like wildfire, fueled by viral videos, deep-fakes, and social media posts that only deepened the mystery. Each new piece of information seemed to contradict the last, leaving a trail of confusion and fear in its wake. The more people dug, the more they uncovered, yet the true purpose behind these actions remained elusive—until now.

This book was born out of a relentless pursuit to understand what lies beneath the surface of this glittering enigma.

It is not just about uncovering who or what is behind the conspiracy, but about understanding the broader implications of these hidden agendas.

Writing this book has been both a journey and a revelation. There were moments when I questioned whether it was worth pursuing, whether the sleepless nights, the cryptic leads, and the dead ends were leading me anywhere meaningful.

But with each new piece of the puzzle that fell into place, I realized that this was more than just a story about glitter; it was a story about power, secrecy, and the lengths to which certain forces will go to maintain control.

As I delved deeper into this tangled web of intrigue, the urgency to uncover the truth became undeniable, as if the future itself hinged on what I might discover..

Be, prepared to see the world through a different lens—one that reveals the connections between seemingly unrelated events, the power plays happening behind closed doors, and the far-reaching consequences of secrets kept from the public eye.

As you turn these pages, you will be stepping into a world where nothing is as it seems, where the mundane is intertwined with the extraordinary, and where every question leads to more questions.

This book is not just an exploration of a bizarre conspiracy; it is an invitation to think critically, to question the narratives presented to us, and to consider the possibility that the truth is often hidden in plain sight.

This isn't just another conspiracy theory to file away with the rest; it's a key to understanding the forces shaping our world.

So, if you've ever wondered whether a faint whisper in the dark could lead to a revelation that blinds with its brilliance, you're about to find out.

Remember, sometimes the truth doesn't just shimmer—it dazzles!

-Justin Time

Chapter1

The Glitter Conspiracy Unveiled

"There's more to the story'

Chapter Question

Could a seemingly harmless material be at the heart of a global power struggle?

As an investigative journalist, I've spent most of my career picking apart hidden truths and shadowy conspiracies. My latest effort has been full of twists, dead ends, and some really unsettling finds. Over the last 18 months I've carefully stitched together a story that makes us rethink what we know about glitter and its potential global impact.

"There are none so blind as those who will not see" –John Heywood

Appearances Are Deceiving

At first, glitter may not seem like a big deal for an exposé. We see it in party decorations, arts & crafts, and makeup. But as I dug deeper, it became clear that there's a lot more than meets the eye. It started simple enough—an odd buying pattern—but quickly turned into something much bigger. You see, this conspiracy could potentially affect every living thing on the planet.

The glitter conspiracy isn't just about sparkle; it's about hidden uses, secret buyers, and maybe even shifting global power dynamics. Some folks think it's vital for advanced military tech, while others say its key for high-tech industries or even something more sinister. But let's back up and see what we actually know first.

"Things are not what they seem. They are what we think they are." – Unknown

What Is the Glitter Conspiracy?

At its core, the glitter conspiracy is about who buys the most glitter. In 2018, journalist Caity Weaver wrote an article for the New York Times titled "What Is Glitter?" She found out that some massive industry was buying tons of it.

The catch? No one would say who or why. This mystery kicked off loads of speculation and led to what's now known as the glitter conspiracy.

What's astonishing is how secretive everything is. Manufacturers and suppliers won't talk, and even people in the industry often don't know who their end customers are or what they do with the glitter. This kind of hush-hush behavior suggests that there's much more going on here than meets the eye.

Exploring the Theories

Any good conspiracy has its theories, right? Here's a few folks are talking about: some plausible, some outlandish, but all worth considering as we take a dive into the deep end of this conspiracy pool.

Military Applications:

There's chatter that glitter is used in stealth tech for cloaking aircraft, ships & soldiers. Its reflective properties might scatter radar signals making detection tougher. Military history shows they use all sorts of unconventional stuff too.

High-Tech Industries: Another good guess? Glitters in optics & electronics. Reflective qualities could improve solar panels or electronic displays. Secrecy might be all about protecting proprietary tech & keeping market advantages.

Currency Production: Some think it's used in making money—literally! Glitter might help create intricate designs on banknotes to thwart counterfeiters.

Cosmetics & Personal Care Products: It's not just sparkly things either; health & environmental concerns are shrouded in mystery too since glitter's a micro-plastic damaging ecosystems (could harm us eventually).

Food & Beverages: Yep! Rumors say some companies use glitter to make products look sparkly but ingesting it? Raises big health risks.

Environmental Impact: With awareness growing around micro-plastic pollution—some claim glitter industries hide their impact to avoid scrutiny/anger from public.

Space Exploration: Sound wild? Maybe! Yet another theory says glitter helps in space—for visibility or insulation purposes on spacecraft!

Covert Operations: Some folks believe it's used by governments for secret missions—reflective optics aiding espionage perhaps?

Art & Fashion: While common here—a hidden agenda by powerful interests creates increasing popularity without clear reasons though...

Mind Control: Outlandish but intriguing nonetheless! Reflective traits inducing hypnotic effects or control behavior adds spice to mystery.

Validating the Conspiracy

Speculation's one thing; proving it's another matter altogether! Throughout my investigation, I spoke with many—government officials, industrial leaders & ex-glitter manufacturer employees. -most wanted anonymity, but off-record chats painted quite vivid pictures, hinting at significant implications due secrets kept tightly guarded.

One ex-employee from a major manufacturer told me they 'strictly prohibited' from asking and questions or making any inquiries about certain customers.

Each employee was individually instructed as to which clients were 'off limits' with explicit warnings against potential fines and possible termination for even discussing them at all! Such secrecy surely means something big.

A high-ranking government official vaguely hinted at military uses yet avoided specifics—indirectly confirming that national security heavily relies on materials like glitter, lending credibility to theories suggesting potential military-dependence scenarios.

"Just because you're paranoid doesn't mean they aren't after you." – Joseph Heller, Catch-22

Summarizing the Glitter Conspiracy

At the heart of this complex web of intrigue is a mysterious, anonymous figure that is secretly purchasing vast quantities of glitter, stockpiling volumes large enough to potentially influence events on a global scale.

The concealed uses of glitter, from defense and technology to cosmetics and security, could influence global dynamics in ways we have yet to fully understand."

In this chapter, we've explored various components of the glitter conspiracy, from its potential military and high-tech applications to its use in currency production and cosmetics. While we haven't provided definitive answers, we've laid the groundwork for further investigation and set the stage for revelations yet to come.

"Conspiracy theories are an irresistible labor-saving device in the face of complexity." – Henry Louis Gates

Conclusion

As we dive into the dark waters of the glitter conspiracy, remember that the truth is often stranger than fiction. The seemingly benign material that we associate with arts and crafts may hold secrets that could change the world. So, as we continue this journey, keep an open mind and be prepared for surprises.

Chapter Question

Could a seemingly harmless material be at the heart of a global power struggle?

Answer

Yes indeed— the glitter conspiracy suggests that this seemingly benign material could indeed be at the heart of a global power struggle, with its secretive applications potentially impacting everything from national security to environmental health

Final Thought to Think About

"The public have an insatiable curiosity to know everything, except what is worth knowing." – **Oscar Wilde**

Chapter 2

Sparkle & Shine: The History of Glitter

"Nothing is as it seems"

Chapter Question

Was glitter's invention a byproduct of military technology development during World War II?

Glitter, with its shiny shimmering allure and ubiquitous presence, seems like a harmless, benign product. However, as we delve deeper into its origins and the industries that surround it, a different, more troubling picture emerges. This chapter reveals the intriguing history of glitter, from its accidental discovery to its rise as a commercial and potentially high-tech asset.

"The past is never dead. It's not even past." - *William Faulkner*

The Discovery

Upon receipt of a formal request from the principals involved, I had to agree to give my word and personal assurance that I would not to identify any of the key players by name. I am thereby bound to omit the name of the inventor and in furtherance of my pledge will not disclose the name of the company.

The origins of glitter are as fascinating as they are complex, and the journey of its creation is deeply intertwined with a very significant historical event. During a pivotal moment in U.S. wartime history, a German-born inventor, whose father was an officer in the German army during World War I, found himself recruited for a top-secret government project. His expertise in micro-precision cutting was crucial to the Manhattan Project, where he played a vital role in developing components essential for the first atomic bombs.

Amid this high-stakes work, he noticed that the byproducts of his borax cutting—tiny, reflective particles—caught the attention and interest of his colleagues. These particles were often scooped up and taken home by his co-workers and used to add sparkle to their birthday celebrations and holiday decorations. This sparked an idea in the inventor, who envisioned a commercial market for this glittering material.

Producing Product

In 1943, he founded a company in New Jersey to turn his vision of creating a commercial market for the shimmering byproduct into a reality, leveraging his expertise and innovative spirit to pioneer an entirely new industry.

He began by constructing a machine in a barn on his family farm to produce and manufacture glitter. Initially, the market for glitter was primarily for decorative and cosmetic purposes and grew slowly.

The inventor was relentless in his effort to develop alternative applications for the product he produced and over time proved successful in expanding the consumer market for glitter.

However, as product applications increased and market awareness expanded, so did demand for the product. Unfortunately for the inventor, this rise in demand led to the rise of competitors.

"Competition is not only the basis of protection to the consumer but is the incentive to progress." – Herbert Hoover

Rivals Emerge

As the inventor's business expanded, he faced growing competition from other glitter manufacturers who seemed to materialize 'overnight, out of nowhere,' entering the marketplace as alternative sources to meet the ever-increasing demand.

He knew that in order to stay ahead, he needed to find a way to produce the product more efficiently and cost-effectively.

He determined that he would have to invent a machine that would create 'zero waste' by converting 100% of the base material into a 'usable product suitable for sale' and thereby be able to produce glitter faster and cheaper than any of his competitors.

Try & Try Again

After countless unsuccessful attempts and month after month of working day and night, dedicating every spare second he could to finding a solution, led to nothing but increased frustration and mounting pressure. The inventor never relented. He was nothing if not tenacious.

It was rumored that he would sing German folk songs out loud to himself as he worked to keep from falling asleep and after a failed attempt he would often joke to his wife that he made "an important discovery... I've discovered something else that doesn't work"

"The greatest test of courage on earth is to bear defeat without losing heart." –
Robert Green Ingersoll

Nature Reveals a Secret

Legend has it that his 'Eureka' moment came while the inventor was taking a bathroom break. He was staring blankly at the tiled floor when he noticed the hexagonal pattern of the tiles. This honeycomb structure, used widely in nature for its efficiency, inspired him.

Excited, he leapt off the toilet with such force that the toilet seat clattered loudly against the porcelain. His mind was racing, adrenaline pumping through his veins as the realization of what he had just uncovered hit him like a lightning bolt.

Without a second thought, he burst out of the bathroom and into the hallway yelling as load as he could "I've got it ... I've got it" His voice, loud and triumphant, echoed throughout the house as he continued shouting at the top of his lungs, "I've figured it out! I've figured it out! I've finally figured it out!"

Full Moon

His wife, who had been peacefully reading in the living room, was startled by the sudden commotion. Her heart skipped a beat as she dropped her book and rushed to see what on earth could be causing such a ruckus. She was halfway down the hallway when she found her husband—whose demeanor was usually so composed, even tempered and rational—lying face down on the cold hardwood floor, laughing uncontrollably.

His pants were tangled around his ankles, leaving him exposed in a way that injected an additional element of brevity into this scene that already appeared to be lifted right out of a situation comedy. It was a detail they would always insist on including whenever they retold the story, as it perfectly captured the absurd nature of this divine moment of inspiration.

A Sparkle of Joy

She hesitated for a moment, unsure whether to laugh, cry, or call for help. But as she watched him, it became clear that his laughter was not one of madness, but of pure, unfiltered joy. This was not the laughter of someone who had lost his mind, but of someone who had just seen the world in a new light, someone who had looked at a problem from every angle and, in the most unexpected moment, found that the answer had been hiding in plain sight.

Finally, catching her breath, she knelt beside him, placing a gentle hand on his back. "What did you finally figure out?" she asked, her voice a mix of curiosity and concern.

He turned his head to look at her, tears of laughter streaming down his face. "You wouldn't believe it," he gasped between chuckles. "I've been wrestling with this problem day after day month after month searching for an answer I was looking at the entire time but couldn't see."

"It's not what you look at that matters, it's what you see." – Henry David Thoreau

Innovation Inspiration

His wife, now smiling, helped him up, and as he stood, still shaky with laughter, they both knew that this was a moment they would never forget—a moment that proved inspiration could strike anywhere, at any time, and that the most profound revelations often come when we least expect them.

This breakthrough led to the development of a die-cutting machine that utilized the structured honeycomb design pattern found in nature, allowing the inventor to produce glitter with little to no wasted base material that wasn't converted to 'product' he normally would have to throw out.

This innovation gave his company a significant competitive edge, enabling it to dominate the manufacturing market by virtue of being able to produce glitter faster and cheaper than anyone else.

"The secrets of nature have often been the foundation of humanity's most groundbreaking innovations." -Unknown

Industry Supremacy

The inventor's innovative spirit didn't end with his first success. Over the years, his company would become the world's largest supplier of glitter. For decades the company would continue to be recognized as the 'Industry Leader' by virtue of its' founder's uncompromising commitment to constant innovation and adaptation.

His son inherited this innovative spirit and made significant contributions, including the invention of holographic glitter. This advancement added another layer of complexity and appeal to the product, further cementing the company's leadership position in the glitter manufacturing marketplace.

"The simple things are also the most extraordinary things, and only the wise can see them." – Paulo Coelho

Roswell, New **Jersey**

The company adapted a strict security protocol that was originally initiated in order to protect their trade secrets and 'intellectual know-how' but was rumored to have been instituted in order to obtain a government security clearance.

It was around this time that the locals started referring to the property around the company's manufacturing facility as "Area 52," a nod to the infamous Area 51, due to its tightened security and the secrecy surrounding its operations.

The invention and evolution of glitter are deeply embedded in a web of intrigue, secrecy, and innovation. From its accidental discovery to its sophisticated applications, glitter's journey continues to reflect the indomitable spirit and relentless pursuit of progress initiated by its inventor.

"A secret's worth depends on the people from whom it must be kept."
— Carlos Zafón

Glitterex and Other Players in the Industry

Founded in 1963, Glitterex has been shrouded in secrecy, with little known about its clients and specific operations. Some claim that Glitterex plays a role in covert government operations, which further complicates the glitter conspiracy.

Adding to the complexity of the glitter industry are international manufacturers who contribute to the global supply of this enigmatic material. In Europe, several companies operate with similar levels of secrecy, supplying glitter for both industrial and cosmetic purposes. These international players often adhere to even stricter confidentiality agreements, which are designed to protect their unique manufacturing techniques and customer lists.

The global nature of the glitter trade means that these companies are competing in a highly secretive and competitive market, where even the smallest leak of information could have significant consequences.

Eco-Friendly Alternative

Moreover, companies like Today Glitter in Miami, Florida, are introducing new dynamics to the industry with their focus on environmentally friendly alternatives. Today Glitter has gained attention for producing biodegradable glitter made from eucalyptus cellulose, addressing environmental concerns associated with traditional plastic-based glitter.

This move towards sustainability is not only a response to growing environmental awareness but also a strategic positioning within an industry that is increasingly under scrutiny for its environmental impact.

As these various players navigate the challenges of maintaining secrecy while responding to changing market demands, the glitter industry remains a curious and closely guarded domain. The intricate web of companies, each with its own secrets and strategies, continues to intrigue those on the outside, adding fuel to the fire of conspiracy theories.

The true extent of glitter's applications—whether in cosmetics, industry, or even defense—remains largely speculative, but the heightened interest and secrecy suggest that the industry's significance may be far greater than it appears on the surface.

Restricted Access

The glitter industry, characterized by a high level of secrecy and protection, employs stringent security measures in order to safeguard its manufacturing and production facilities.

The manufacturing plants are heavily guarded and access to customer information is tightly controlled. Employees are threatened with 'severe consequences' and harsh repercussions' if they disclose 'anything at all' about certain clients.

These 'warnings' are veiled threats meant to intimidate the employees and are often listed in specific detail as damages for violations of their legally binding 'Non-Disclosure' (NDA) confidentiality agreements. This heightened secrecy only fuels speculation and helps foster the spread of conspiracy theories regarding the true purpose and potential uses of glitter.

"The power of secrecy is immense, and what one is ignorant of becomes the mystery and magic of another." – William Thackeray

Conclusion

Glitter's history is as reflective and multifaceted as the product itself. The journey of glitter from a byproduct of the Manhattan Project to a commercial success is a tale of innovation, secrecy, and potential global intrigue. The continued dominance of manufacturing by two companies located in New Jersey only add to the mystery.

The absence of answers invites us to begin to question the true nature and potential implications of this shimmering material.

Chapter Question

Was glitter's invention a byproduct of military technology development during World War II?

Answer

Yes, glitter was born out of the military technology development during World War II, specifically as a byproduct of the Manhattan Project. It played a crucial role in the development of the atomic bomb, and its journey from a military byproduct to a commercial product is shrouded in secrecy and intrigue.

Final Thought to Think About

"Innovation is born from necessity, but its journey is often paved with unexpected twists and turns."

- Steve Jobs

Chapter 3

Glitter's Market Disparity

"Follow the Money"

Chapter Question

Who buys all the glitter and why is there such secrecy surrounding its customers?

Glitter, with its dazzling appearance and seemingly innocent applications, has a market that is as mysterious as it is expansive. The secrecy surrounding its customer base and the sheer volume of glitter produced and sold annually has given birth to a multitude of conspiracy theories. Who buys all the glitter? Why is there such a shroud of secrecy over its consumers?

"The secret to creativity is knowing how to hide your sources." – Albert Einstein

The Glitter Market Overview

Glitter? It's big business, worth billions! Uses range from arts & crafts to high-tech and industrial uses. The industry's top two Major manufacturers, both located in New Jersey, are considered 'the big players' that run the show.

But here's the twist—despite the massive amounts being sold, there's a big gap between what's produced and the amount accounted for in consumer and commercial applications.

According to a report from Global Market Insights, the glitter market was valued at approximately USD 4.9 billion in 2020 and is projected to grow at an annual rate of 5.2% from 2021 to 2027.

The majority of this market is driven by the cosmetics industry, followed by arts and crafts, industrial applications, and others. However, a considerable portion of the glitter produced remains unaccounted for, fueling speculation and giving rise to conspiracy theories.

"From a tiny spark may burst a mighty flame." – Dante Alighieri

Consumption Discrepancy

In 2021, it was reported that over 63% of the glitter produced globally was unaccounted for in consumer applications. This discrepancy has only increased in subsequent years, with 2022 and 2023 showing similar trends.

The question arises: where is all this glitter going? The mystery deepens when we consider the high level of secrecy maintained by the major producers and their clients.

One significant piece of evidence comes from a 2021 report by Glitterex, which revealed that a large, undisclosed client purchases an enormous volume of glitter annually.

This client, shrouded in secrecy, has a standing purchase order that accounts for a 'rather substantial portion' of Glitterex's overall production.

Despite numerous inquiries, neither Glitterex nor the other leading producer would disclosed any information regarding the identity of this 'mystery' client or even provide a single clue concerning the intended use and application of this 'very large volume' of warehoused glitter.

Industrial and High-Tech Applications

While glitter is commonly associated with arts and crafts, its applications extend far beyond that. In the industrial sector, glitter is used in various manufacturing processes, including automotive paints, industrial coatings, and reflective materials. Its reflective properties make it invaluable for safety applications, such as road markings and emergency signage.

In the high-tech sector, glitter is used in advanced materials and electronics. Its unique optical properties make it useful in the deployment of anti-counterfeit measures for currency and documents, ensuring security and authenticity in financial transactions.

Additionally, glitter is employed in the production of holographic displays and other advanced visual technologies, elevating the impact of special effects and enhancing the realism of 3D displays. However, these applications do not account for even a fraction of the significant volume of glitter that remains unexplained, Who is purchasing it and why???

"The less we know, the more we suspect." - Josh Billings

The Mystery Buyer

Who could this mystery buyer be? The secret identity of the mystery buyer remains one of the most intriguing aspects of the glitter conspiracy. Speculation ranges from government agencies to large multinational corporations.

The most popular 'Glitter Conspiracy' theories suggest that the buyer is involved in some covert government operation and intends on using glitter for a purpose that is related to national security that has been 'classified' as 'Top Secret.'

Justifiable Concern or Paranoia

I had a somewhat strange interview with a former employee of Glitterex that insisted I first forward him with a 'legally binding' signed statement swearing that his identity would remain anonymous that his lawyer had to approve before he could agree to talk to me.

During the course of our nearly two hour conversation that I can only describe as 'highly uncomfortable and most unusual' he revealed that there was no length the company would not go to in order to maintain the secrecy and confidentiality of their biggest client.

"We had strict instructions never to discuss this particular client or talk about anything that had to do with their orders under fears that included penalties and large fines along with the possibility of 'immediate termination," the seemingly still frightened former employee stated. "

In addition "there were numerous NDAs in place that specifically addressed this issue, and the level of secrecy was akin to working at a top-secret government agency like the CIA," he further explained, adding to the mystery and intrigue.

"Secrecy is the enemy of efficiency, but it is also the tool of control." – David Brin

Environmental and Health Concerns

Another angle to the glitter conspiracy is the environmental and health concerns associated with its production and use. Glitter is essentially a micro-plastic pollutant, and its widespread use has raised alarms about its potential impact on the environment. Glitter is considered 'forever products' known to persist in the environment for hundreds of years, posing a significant threat to marine life and related ecosystems.

Studies have shown that micro-plastic pollutants, including glitter, have been found in the water supply, soil, and digestive systems of marine organisms. Micro-plastics pollutants have also recently been detected in the human blood stream. This finding was 'alarming' and initially raised a high level of concern due to the inability of the body to expel this contaminate – once it gets inside, it stays inside!!

Environmental Impact Ignored

But despite having been identified as a 'major contributor ' and 'one of the leading sources' of micro-plastic pollution, the potential dangers posed by glitter, were all but ignored by the mainstream media and largely dismissed by the scientific community.

The secrecy surrounding the glitter industry has fueled suspicions and has given rise to a widely held conspiracy that believes that the true extent of its environmental impact is being deliberately concealed. This concealment aims to maintain a positive public perception of glitter, ensuring its continued widespread use and acceptance despite the extent of the potential ecological damage it can cause, once again contributing to the mystery of 'who & why?'

Government and Industry Secrecy

The glitter industry's high level of secrecy has not gone unnoticed. Official Government agencies and regulatory bodies have been scrutinizing the industry over the past several years, but with very limited success.

This lack of transparency has only added to the intrigue, with some speculating that government agencies themselves are involved in the covert use of glitter.

The secrecy also extends to trade shows and industry conferences, where glitter manufacturers are only accessible via closed-door meetings that are held in restricted no-access areas.

Journalists and researchers have reported incredible difficulty in obtaining any information or even basic access to these events, further fueling speculation about the intended agenda hidden behind the glitter conspiracy.

"Governments constantly choose between telling lies and fighting wars, with the end result always being the same. One will always lead to the other." – Thomas Jefferson

'Stop' Signs

I have encountered numerous roadblocks and unforeseen challenges over the past year and a half working as an investigative journalist as I gathered the evidence needed in order to solve the mystery behind the glitter conspiracy.

My multiple attempts to tour the 'two major manufacturing facilities located in New Jersey' were met with harsh resistance and vague explanations.

"Truth is like the sun. You can shut it out for a time, but it ain't goin' away." –

-Elvis Presley

No Trespassing

On one occasion, I was even physically escorted off the premises by security and told that if I returned 'unannounced' and 'uninvited' I would be immediately arrested for trespassing. This only served to reinforce my suspicions and strengthen my determination to find the answer as to why such a high degree of secrecy surrounds the daily operations of these companies.

Conversations with industry insiders and former employees often ended abruptly when the topic of the mystery buyer was brought up. One former employee, who also insists on remaining anonymous, hinted at a possible large scale foreign government connection. "Let's just say that our biggest client is not who you'd expect," they said, "And their intended use will directly affect every person on the planet."

Speculation and Conjecture

The glitter conspiracy has given rise to numerous theories and speculations. Some believe that glitter is being used in advanced military applications, such as stealth technology or electronic signal jamming.

Others suggest that it is a critical part of a larger plan that seeks to control the global economy through covert operations and market manipulation.

One particularly intriguing theory is that glitter is being used in secret space missions. The thermal distribution and reflective properties of glitter could make it useful for various applications in space, as a heat-shield insulating agent, a rocket fuel additive or even as an anti-radiation containment apparatus.

Support for this theory points to the high level of secrecy and the direct and indirect involvement of government agencies within the glitter industry.

"If you shut up truth and bury it underground, it will but grow."

- Emile Zola

Conclusion

The glitter conspiracy is composed of a complex web of secrecy, speculation, and intrigue. The significant discrepancy in the volume of glitter produced and sold compared to the amount of glitter that enters the marketplace.

While the mystery buyer remains unidentified, the level of secrecy surrounding glitter's biggest customer raises important questions about its true and intended purpose. The. potential implications of the 'glitter conspiracy' are far-reaching and mind boggling. The secrecy and intrigue surrounding the glitter industry suggest that there is 'much more to this story than meets the eye.'

"All truths are easy to understand once they are discovered; the point is to discover them."   - Galileo Galilei

Chapter Question

Who buys all the glitter and why is there such secrecy surrounding its consumers?

Answer

The glitter industry is shrouded in secrecy, with a significant portion of its production unaccounted for in consumer applications. The mystery buyer, whose identity remains undisclosed, purchases a substantially large volume of glitter, fueling speculation and conspiracy theories. The potential implications of this conspiracy are profound, suggesting connections to covert government operations, advanced military applications, and global power dynamics.

Final Thought to Think About

"Truth is mighty and will prevail. There is nothing the matter with this, except that it isn't so" **–Mark Twain**

Chapter 4

Historic Validation of Conspiracies

"It's all connected"

Chapter Question

Are conspiracy theories mere fantasies, or do they often contain kernels of truth?

Conspiracy theories have been part of our story for ages, driven by our curiosity and doubt. They spring up from a mix of uncertainty, secrecy, & a lack of trust in those in charge. Though many get brushed off as nonsense, history shows that some carry a grain of truth, hidden under layers of secrecy. In this chapter, we investigate significant conspiracies, uncovering the proof that indicates things may not be as they appear."

"Just because you're paranoid doesn't mean they aren't you. — *Joseph Heller*

The Reality of Conspiracies in History

Conspiracies have been part human history for as long as there been societies. From empires to modern states, the allure of hidden agendas and secret plots has always captured the public's imagination. Many conspiracy theories are speculative, but history shows that some conspiracies are rooted in truth. This chapter digs into the historical validation of conspiracies, especially those that have shaped the United States.

The term "conspiracy theory" often brings to mind crazy, unfounded accusations. But really, some big events in history involved real conspiracies. Take the Watergate scandal, for instance; it seemed wild at first but eventually led to President Nixon stepping down. Then there's the Iran-Contra affair, which showed a tangled web of secret operations by high-ranking officials.

In the U.S., suspicion & mistrust have grown thanks to many revelations exposing the shady sides of political & corporate power. During the Cold War, secret government actions, intelligence agency shenanigans, and sneaky deals by big companies fueled this paranoia.

Puppet Masters

Lots of people think unseen forces are pulling strings behind the scenes—and honestly, it seems pretty likely. Look at the infamous MK-Ultra program. People thought it was just paranoid nonsense at first. But later, we found out that the CIA really did conduct mind control experiments on unsuspecting folks. Stuff like this inflamed beliefs that governments would go to extreme lengths for their goals.

The Tuskegee Syphilis Study is another terrifying example. For years, African American men were tricked and denied proper medical care so researchers could watch syphilis progress. This terrible secret weighed heavily on the minds of many African Americans—leading to deep mistrust in medical & governmental institutions.

These stories remind us that while not every conspiracy theory is true, some certainly are. And it's these validated conspiracies that make us question what else might be lurking in the shadows.

"When it comes to the government and their plans, the less you know, the better."

– Frank Zappa

Public Distrust & Suspicion

The climate of the 21st century, filled with rapid tech changes & growing government snooping, has only boosted the belief that 'they are hiding something from us' or 'not telling us the truth', leading to an ever expanding distrust of government entities and institutions.

Whistleblowers like Edward Snowden have shown just how far governments are willing to go to monitor and control their citizens. The NSA's huge spying programs, once viewed as 'unfounded paranoia' or the foundation of a futuristic sci-fi movie, have been confirmed as being 'the truth'. (for real for real!)

Shaping Perception

Corporate conspiracies have also played a major role in shaping the public's perception. Remember how the tobacco companies hid just how dangerous smoking is? They knew it was bad but kept it quiet. This is a stark reminder of how corporate interests can overshadow public health. Fossil fuel companies? They've been accused of playing down carbon emissions' impact on climate change just to protect their profits.

Now, the glitter conspiracy might seem silly and trivial at first glance, but it fits into this historical pattern of secrecy and deception. The secrecy around who uses glitter and why makes you wonder what's really going on. Looking at history, we see it's hard to tell the difference between a conspiracy theory and a conspiracy that has been proven to be true.

"A conspiracy theorist is someone who questions the statements of known liars." – Unknown

A History Of Conspiracy

The U.S., with its unique culture and politics, has always been fertile ground for conspiracy theories. Even the nation's founding itself was orchestrated by means of secret plots & underground meetings.

Take the Boston Tea Party—secretly planned by the Sons of Liberty, was a pivotal event leading up to the American revolution. This tradition of subterfuge has continued throughout American history, from the assassination of President Lincoln to the covert operations of the CIA.

In today's modern era, the internet has proven to be a powerful tool that is quite proficient at promoting—and debunking — conspiracy theories.

Online communities provide a platform for like-minded individuals to share information, speculate as to what it means and what it means they're not telling us. Sometimes they promote paranoia but other times they find out the truth! But these same sites can spread misinformation and distort the facts to fit a narrative, making it tough to know what's real and what's not.

"The essence of government is power; and power, lodged as it must be in human hands, will ever be liable to abuse."
— James Madison

What History Has Taught Us

As we identify and examine examples of validated conspiracies in this chapter, it's essential that you keep an open mind. While skepticism is necessary, history has shown that some conspiracies are not 'theories' but are in fact 'facts'.

The tricky part is knowing the difference between what can be and what has been proven to be true with what could be found to be true but has yet to be proven. Another real challenge is understanding the drive, motivation and purpose behind those that are .seeking the truth and those who seek to keep the truth hidden.

Next up, we'll shed some light on the mechanisms of secrecy and control that have shaped our world by examining some notable conspiracies that have been proven true. By learning from these historical precedents, we can better understand and appreciate the complexities of the glitter conspiracy and the potential implications for what it might mean.

"The very word 'secrecy' is repugnant in a free and open society." – John F. Kennedy

The 9/11 Attack & Saudi Government Involvement

The events on September 11, 2001, rocked the world & started a global war on terror. Almost right away, people began doubting the official story from the U.S. government.

One strong theory indicated that the Saudi government was somehow directly involved and they provided the terrorist with the financing and funding needed to coordinate and carry out the attack.

Even though the 9/11 Commission Report from 2004 didn't find solid proof linking the Saudi government to the attacks, secret documents and stories from different sources kept this idea alive. In 2016, the release of the "28 pages," a secret part of the report, hinted at a connection between Saudi nationals and the hijackers.

But still, the U.S. government, citing national security, played it safe and denied there was any evidence that supported the theory that the Saudi Government financed or was directly involved in the attack. The theory about the Saudi Government's role in 9/11 gains strength from actions like those by Omar al-Bayoumi, who had direct ties with two of the hijackers.

People closely connected to the incident still believe the government has been deliberately withholding information from being released because it will prove unequivocally that the Saudi Government not only knew about the planned attack, they financed it!

The Assassination of Journalists and Dissenters

The shocking murder of journalist Jamal Khashoggi in 2018 made headlines worldwide and showed how far some will go to silence critics. Khashoggi was a known critic of the Saudi government & was killed inside their consulate in Istanbul.

At first denied by Saudi officials, they later admitted it under pressure once a video surfaced showing Khashoggi entering the consulate but never leaving.

This isn't the first time such things happened. We've seen similar state-sponsored killings before; remember Alexander Litvinenko poisoned in London or Sergei Skripal attacked with nerve agent in Salisbury?

Such cases add to patterns of secrecy &denial, offering proof that these conspiracies are rooted in real events.

"The lie can be maintained only for such time as the State can shield the people from the political, economic and/or military consequences of the lie." – Joseph Goebbels

The Assassination of JFK, RFK, and MLK

The '60s were noisy times for America—full of turmoil—marked by JFK's assassination (1963), RFK's assassination (1968), & MLK Jr.'s assassination (1968). Each tragic event sparked numerous conspiracy theories; many Americans doubted what they were officially told.

John F. Kennedy: in '63 the Warren Commission determined that Lee Harvey Oswald acted alone. But conflicting evidence & the mysterious deaths of multiple witnesses has kept alive speculations about a larger conspiracy involving the CIA, LBJ or even the Mafia.

Robert F. Kennedy: When RFK died by Sirhan Sirhan's hand people thought powerful interests were behind his death because he opposed Vietnam War & had a high chance of winning the presidency.

Martin Luther King Jr.: James Earl Ray confessed but later took it back saying he got set up! King family even believes Ray wasn't guilty claiming bigger forces were at play that aimed to silence MLK because his voice had grown too loud.

The UFO Phenomenon

Unidentified Flying Objects (UFOs) have captivated public imagination for decades, with numerous sightings and encounters reported worldwide. The U.S. government's handling of UFO information has been marked by secrecy and denial, fueling conspiracy theories about extraterrestrial life and advanced technology.

The Roswell incident in 1947 is perhaps the most famous UFO case. The initial announcement of a recovered "flying disc" was quickly retracted, with the military attributing the debris to a weather balloon. Decades later, declassified documents revealed the existence of Project Mogul, a top-secret program involving high-altitude balloons. However, many still believe the Roswell incident involved an extraterrestrial craft.

In recent years, the U.S. government has taken a more transparent approach, acknowledging the existence of the Advanced Aerospace Threat Identification Program (AATIP) and releasing videos of unidentified aerial phenomena (UAP).

"The best way to keep a secret is to pretend there isn't one." – Margaret Atwood

The Existence of the Deep State

The concept of a "deep state" refers to a clandestine network of powerful individuals and institutions that operate independently of elected officials to influence government policy and decisions. While often dismissed as a paranoid fantasy, historical events and leaked documents suggest that elements of a deep state may actually exist.

The Watergate scandal in the 1970s exposed the extent of political corruption and surveillance within the U.S. government. Similarly, the Iran-Contra affair revealed covert operations carried out without congressional approval. These incidents demonstrate that secret, shadowy networks that are hidden deep within the government can and do operate without public knowledge and without public scrutiny.

The release of classified documents by whistleblowers such as Edward Snowden has further fueled the belief in the existence of a deep state. Snowden's revelations about mass surveillance programs conducted by the National Security Agency (NSA) highlighted the extent to which the government will go in order to monitor and control your personal information.

Government and Industry Secrecy

Throughout history, the government and powerful industries have often colluded to keep information from the public. The Tuskegee syphilis experiment, which ran from 1932 to 1972, involved the U.S. Public Health Service conducting unethical medical experiments on African American service men without their knowledge or informed consent.

This scandal, along with others like the CIA's MKUltra mind control program, illustrates the lengths to which authorities will go to hide and conceal the truth from the public.

The secrecy surrounding the glitter industry, with its mysterious buyer and undisclosed application, fits right into this historic pattern of government and industry collusion. Despite repeated denials and obfuscations, the persistence of secrecy surrounding this industry raises legitimate questions about exactly 'what' is being hidden from the public and more importantly 'why?'

"A nation that is afraid to let its people judge the truth and falsehood in an open market is a nation that is afraid of its people." – John F. Kennedy

Dance the Dance

Conspiracy theories often arise from a blend of fact and fiction, with the truth buried somewhere under multiple layers of secrecy and denial. The historical validation of conspiracies such as the 9/11 attacks, state-sponsored assassinations, UFO phenomena, and the existence of a deep state demonstrates that there is often much more to these theories than what initially appears.

The intricate dance of information and misdirection serves as a reminder of the complexities involved in groundbreaking endeavors. As we piece together the puzzle, each fragment of truth brings us closer to understanding the full scope of the glitter conspiracy. The revelation that Musk's clandestine mission could be our last, best hope against climate change redefines the narrative from one of fear to one of cautious optimism.

Historical Reflections

In reflecting on these historical conspiracies, it becomes clear that the mechanisms of secrecy and control are deeply ingrained in the structures of power. The glitter conspiracy, with its layers of mystery and intrigue, fits into this long-standing pattern.

Understanding the history of validated conspiracies helps us recognize the potential reality behind seemingly far-fetched theories and underscores the importance of vigilance and skepticism in our search for the truth.

As we continue to unravel the glitter conspiracy, we must remain aware of the lessons from history, recognizing that behind every veil of secrecy, there is often a kernel of truth waiting to be discovered. This chapter has provided a foundation for understanding the broader context of government and industry secrecy, setting the stage for the revelations to come.

"History is a set of lies agreed upon." –
Napoleon Bonaparte

Conclusion

In the following chapters, we will delve deeper into the specifics of the glitter conspiracy, examining the key markets, hidden applications, and the potential implications of this enigmatic phenomenon. By keeping an open mind and continuing to question the official narratives, we can uncover truths that they kept hidden and expose the intricate web of deception they have woven to keep the public trapped in darkness.

Chapter Question

Are conspiracy theories mere fantasies, or do they often contain kernels of truth?

Answer

Historical evidence suggests that many conspiracy theories, while often dismissed as baseless, often contain hidden kernels of truth. The glitter conspiracy, with its significant market disparity and mysterious buyer, fits right into this historical pattern observed in these conspiracies of secrecy, government collusion, and aggressive denial and validates the need to remain skeptical and vigilant.

Final Thought To Think About

"The truth is like a lion; you don't have to defend it. Let it loose; it will defend itself."
– Saint Augustine

Chapter 5:

Noncommercial & High-Tech Applications

"Question everything

Chapter Question

Could glitter be the secret ingredient behind groundbreaking technological advancements and state-of-the-art applications?

In the vast landscape of conspiracy theories, certain threads weave together to create a tapestry of intrigue and speculation. One such thread is the role of glitter in non-commercial and high-technology applications. This chapter will explore these and provide insights into how glitter's unique properties make it an invaluable asset in various industries.

"The science of today is the technology of tomorrow" *– Edward Teller Lang*

The Technology Conspiracy

Throughout history, many conspiracy theories have emerged from the shadows, revealing kernels of truth that were once hidden from public view. The glitter conspiracy is no different.

As we delve into the high-tech and non-commercial uses of glitter, we will see how this seemingly innocuous material plays a crucial role in advanced technologies and government projects. This connection adds another layer of complexity to the glitter conspiracy, raising questions about what else might be concealed from the public eye.

The glitter conspiracy is not just about the sparkling particles we see in art and crafts. It involves the use of glitter in ways that most people would never imagine.

From enhancing the security features of currency and identifying original documents to improving the performance of electronic devices, glitter has found its way into some of the most sophisticated technologies of our time.

"The advancement of technology is based on making it fit in so that you don't really even notice it, so it's part of everyday life." – Bill Gates

Advanced Application Sensation

While glitter may seem like a harmless decorative material, its unique properties have made it an essential component in various industries, from anti-counterfeit measures to advanced optical technologies.

In the world of anti-counterfeit measures, glitter is used to create holographic images and other security features that are difficult to replicate. This helps protect against forgery and fraud, ensuring the integrity of important documents and currency.

However, the use of glitter in these applications raises questions about who controls this technology and what their true intentions might be. Could there be a hidden agenda behind the widespread use of glitter in anti-counterfeit measures?

In the realm of advanced visual technologies, glitter is used in the creation of holographic displays and other cutting-edge optical devices. These applications take advantage of glitter's ability to manipulate light and create stunning visual effects. But again, we must ask ourselves: what else might glitter be used for that we are not yet aware of?

Making the Impossible Possible

Glitter's unique properties also make it useful in electronic devices, where it can enhance performance and efficiency. From improving the conductivity of materials to reducing energy consumption, glitter has found its way into some of the most advanced electronic technologies. This raises further questions about the true purpose of incorporating glitter into these applications and whether or not there might be hidden agendas at play.

"The only way to discover the limits of the possible is to go beyond them into the impossible." – Arthur C. Clarke

Who, Why & How

Throughout this chapter, we will present various examples of non-commercial and high-tech applications of glitter, drawing on source materials whose references have been included to easily enable additional inquiries into this enigmatic material. We will also explore the potential implications of these applications and how they 'dove-tail' into the broader context of the glitter conspiracy.

By keeping an open mind and by continuing to question the official narratives we may be able to uncover truths that those in power would prefer to keep hidden.

As we delve deeper into the specifics of the glitter conspiracy, we will examine the main players involved in these high-tech applications. Who is in control of technologies that depend on glitter, and what are their long term objectives? Are there any hidden agendas that we should be aware of, or is there an innocent explanation for the widespread use of glitter in high-technology applications?

"He who has a why to live can bear almost any how." – Friedrich Nietzsche

Question Everything

The information presented in this chapter is intended to facilitate further research and encourage readers to explore these topics in greater detail. By closely examining the source materials and objectively considering the evidence, we can gain a better grasp of the depth of the glitter conspiracy and its potential implications.

Dive Into The Deep End

This chapter will provide a foundation for this exploration, offering insights and perspectives that will challenge the credibility of the official narratives.

Be prepared to look beyond the surface and come along on this deep dive as we search for the 'specifics'. We will explore the various industries and applications where glitter is used, laying out the facts and backing them up with evidence.

By chapter's end, you'll begin putting the puzzle pieces together that paint a clearer picture of the breadth of the glitter conspiracy and its potential impact. see a clearer picture of the get a clearer picture of the glitter conspiracy & its stakes.

Be ready to question everything you think you know and all of what you've been told as we continue on our quest to unearth the truth behind the glitter conspiracy.

"Question everything. Learn something. Answer nothing." – Euripides

Glitter in Optics & Photonics

First off, let's talk about optics and photonics. Glitter's reflective abilities make it ideal for these fields that rely on light manipulation. It enhances optical devices big time.

Solar Panels:

Researchers are hard at work using glitter-infused solar panels to increase efficiency. Adding glitter to photovoltaic enables it to capture more sunlight and convert it into energy more effectively.
Source: "Glitter in Solar Panels" – Journal of Photonic Research, 2022.

Holographic Displays:

Glitter's holographic features are game-changers for making advanced displays with vivid, realistic images. This could shake up entertainment, ads, & medical imaging.
Source: "Holographic Displays and Glitter" – Optics and Photonics News, 2021.

Medical Applications

Even the medical field sees potential in glitter. Its distinct properties help develop new diagnostic tools & treatments.

Diagnostic Tools:

Glitter particles are used in certain biosensors to detect minute concentrations of biomarkers, allowing for earlier and more precise disease detection.

Source: "Biosensors and Glitter" – Medical Innovations Journal, 2022.

Targeted Drug Delivery:

Scientists are also looking at using glitter in targeted drug delivery systems. Glitter particles can be engineered to carry medicine straight to sick cells, minimizing side effects while improving treatment efficacy.

Source: "Targeted Drug Delivery with Glitter" – Journal of Biomedical Engineering, 2023.

Environmental Monitoring

Glitter's reflective and refractive properties are being utilized in environmental monitoring technologies, providing new ways to track and analyze environmental changes.

Water Quality Monitoring:

Sensors with glitter particles spot pollutants and contaminants with high precision. This helps ensure safe water quality.

Source: "Glitter-Based Water Quality Sensors" – Environmental Science & Technology, 2022.

Climate Monitoring:

Satellites for climate monitoring are getting an upgrade with glitter too. The reflective properties of glitter help improve the accuracy of satellite-based measurements of atmospheric conditions.

Source: "Glitter in Climate Monitoring" – Journal of Atmospheric Science, 2021.

High-Performance Materials

The unique physical and chemical properties of glitter are being leveraged to develop high-performance materials in various industries.

Heat-Resistant Coatings:

In aerospace & automotive industries, heat-resistant coatings made with glitter can withstand super high temperatures.

Source: "Heat-Resistant Coatings with Glitter" – Materials Science and Engineering, 2023.

Wear-Resistant Materials:

Industrial machinery benefits from wear-resistant materials infused with glitter—they last longer and need less upkeep.

Source: "Wear-Resistant Materials and Glitter" – Journal of Industrial Engineering, 2022.

Communication Technologies

Glitter's ability to manipulate light is also being explored in the field of communication technologies, where it could play a crucial role in the development of next-generation fiber optic networks.

Fiber Optic Cables:

Researchers working with fiber optic cables enhanced with glitter transmit data faster & cut down signal loss.

Source: "Glitter in Fiber Optics" – Communications Technology Review, 2022.

Li-Fi Technology:

Li-Fi (Light Fidelity) tech uses light for data transmission. Add some glitter & it becomes more efficient—including a broader range than traditional Wi-Fi.

Source: "Li-Fi and Glitter" – Journal of Wireless Communication, 2021.

Highway Safety

The automotive industry has also found innovative uses for glitter as well, particularly in the development of self-driving cars and increased highway safety.

Highway Signs and Lane Markings:

They're making road signs and lane markings easier to see by adding glitter. This helps self-driving cars navigate roads safely.

Source: "Glitter in Highway Safety" – Journal of Automotive Safety, 2022.

Electric Car Batteries:

Glitter particles are being explored as a component in the development of more efficient electric car batteries. These batteries offer uniform energy storage and enhanced performance, contributing to the production of longer-lasting batteries and more reliable electric vehicles.

Source: "Glitter in Electric Car Batteries" – Journal of Energy Storage, 2023.

Artistic & Creative Applications

Of course, don't forget the arts! In addition to its scientific and industrial uses, glitter continues to inspire creativity and innovation in the arts and entertainment industry

.Interactive Art Installations:

Artists use glitter in exterior art installations that react to changes in light/movement—creating cool, dynamic experiences for viewers.
Source: "Glitter in Interactive Art" – Art and Technology Magazine, 2022.

FFAugmented Reality (AR) and Virtual Reality (VR):

Reflective properties of glitter enhance AR/VR experiences to be more immersive—great for gaming or education!
Source: "Glitter in AR and VR" – Virtual Reality Research,2023

> ***"Truth is stranger than fiction, but it is because Fiction is obliged to stick to possibilities; Truth isn't."*** – Mark Twain

Conclusion and Reflection

The noncommercial & high-tech stuff around glitter show just how versatile it is. It's amazing. From better medical stuff to top-notch environmental monitoring and wild new ways to communicate, glitter's uniqueness is really changing the game. But hey, as we cheer these cool advancements, let's not forget there are deeper issues hiding behind all that sparkle.

Economic Implications

Glitter is a big deal money-wise too. Companies that get how to use it right can win big, shaking up entire industries. Imagine solar panels with glitter making renewable energy cheap! Awesome, right? Or think about faster internet with glittery fiber optics boosting the tech world & creating new jobs.

Environmental Concerns

Sure, glitter helps with environmental monitoring, but making & throwing it away isn't easy. It's tough stuff that can stick around and hurt ecosystems. Researchers need to create an eco-friendly biodegradable alternative option.

Health Impacts

Using glitter in medicine could be game-changing for diagnosing and treating diseases. But we don't really know what long-term exposure does, especially tiny bits of it (nano-glitter). More studies are needed to figure out if it's safe overall in healthcare use.

Technological Advancements

Glitter's a game-changer in things like fiber optics, Li-Fi (that's light-based Wi-Fi), and electric car batteries. These boosts can transform industries and improve our daily lives big time. For instance, faster internet or longer-lasting electric cars all thanks to the magic of glitter!

Mystery and Secrecy

The secrecy about what glitter's used for makes people suspicious. Who is buying and stockpiling tons and tons of this shiny stuff?

What are they up to? The lack of transparency fuels rumors and suggests that it may involve the use of glitter in a high tech commercial or military application.

Potential for Abuse

Any strong tech can be misused. Glitter's no different—the awesome properties could be twisted for bad stuff. Think spy devices again: yikes! That's why rules are vital to stop misuse and protect rights while enjoying the benefits.

The Role of Government

The government stepping in can be reassuring but also worrying due to control issues. While oversight ensures safety & ethics, it raises questions about who holds the power over what?

Future Prospects

It's thrilling but uncertain where glitter tech heads next! Ongoing research will likely reveal many more uses but must balance progress with ethical considerations carefully.

Public Awareness

People need to know what is going on with high-tech glitter so there can be discussions about the responsible development of applications without having to compromise any public safety.

"The possession of knowledge does not kill the sense of wonder and mystery. There is always more mystery." – Anaïs Nin

Surprised?

As we made our way through this chapter it became clear that glitter ain't just another 'flash in the pan'. Its uses extend across many industries—from cosmetics and arts& crafts to cutting-edge technologies & even military applications.

Every use hints at there being something bigger behind it, suggesting glitter's role might be way more important than it would initially appear.

When I started investigating the glitter conspiracy, I never imagined the depth and breadth of its potential applications.

What surprised me the most? The high-tech stuff, like anti-counterfeit technology and holographic displays. These uses? Super cool! But they also make me wonder just how much glitter is woven into our daily lives? What are the broader implications of its intended use? And what might that mean for the rest of us??

"There is nothing more deceptive than an obvious fact." – Arthur Conan Doyle

Can't Get It Out Of My Head

One thought that kept bouncing around in my head was what about glitter's impact on the planet? It is a 'forever product' that doesn't go away and the possibility that its true environmental impact is being deliberately concealed are areas concern me greatly.. r doesn't just disappear—it sticks around & could cause major environmental problems.

And get this—the real environmental impact might be hidden from us on purpose! It's not just about the sparkle; it's also about what it is doing to the Earth and its ecosystems and what that means to the health of current and future generations.

I also couldn't stop wondering why such a high level of secrecy surrounds the glitter industry. The tight control over information, the guarded production facilities, and the reluctance to disclose client lists all point to something that is being kept from us that we very well might need to know. This secrecy fuels suspicion and makes you really start to wonder who is keeping what from us and why?

"Who controls the past controls the future. Who controls the present controls the past."
– George Orwell

Still Puzzled

During my research, I found that the more that I dug for answers the more questions I unearthed. Solving the glitter conspiracy feels like trying to put a puzzle together that has pieces missing that have been scattered all over different industries and applications.

Each piece reveals a bit more of the picture, but the more pieces you find the more pieces you need and unless and until you 'connect the dots' you won't be able to figure out what you're looking at!!!

One really troubling thought is the idea that the government might be directly involved in concealing the intended 'large scale' targeted application of this micro-plastic.

'Given the history of government secrecy and media manipulation, it's not too far-fetched an idea. The potential for glitter to be used in military applications or other covert defense operations only adds to the intrigue and increases suspicion.

"Secrecy, being an instrument of conspiracy, ought never to be the system of a regular government." – Jeremy Bentham

Connecting The Dots

As we continue to unravel the glitter conspiracy, it's important that we remain undaunted and determined to solve this mystery. to stay alert and skeptical. The pieces of the puzzle are slowly coming together, but there are still too many unknowns. The connections that tie together glitter, advanced technology, government secrecy, and potential environmental impacts are all threads that need further examination.

"The first step toward change is awareness. The second step is acceptance." -Branden

Conclusion

In the next chapter, we'll delve even deeper into the details that define the glitter conspiracy, examining the key players, hidden agendas, and potential implications of this enigmatic phenomenon. By keeping an open mind and continuing to question the official narratives, we may be able to shed some light on the truth and expose answers, information and actions that those in power hid in the shadows in order to keep us in the dark!

Chapter Question

Could glitter be the secret ingredient behind groundbreaking technological advancements and state-of-the-art applications?

Answer

Yes, glitter's unique properties have found applications in various high-tech fields, from optics and photonics to medical diagnostics and environmental monitoring. However, these advancements come with significant ethical, environmental, and health considerations that must be addressed to ensure responsible development and use.

The secrecy surrounding glitter's applications further fuels speculation and underscores the need for transparency and accountability.

Final Thought To Think About

"Any sufficiently advanced technology is indistinguishable from magic."

– Arthur C. Clarke

Chapter 6

Glitter and Military Defense

"The government is hiding something"

Chapter Question

Could glitter be the secret weapon that turns the tide in modern warfare?

When we think of glitter festive decorations and shimmering beauty products come to mind. However, the reality is far more complex and potentially alarming. Glitter's unique properties make it a valuable asset in military and defense applications. This chapter delves into the various ways glitter might be employed in military operations, examining both the science and the speculation.

"In times of war, the simplest materials can become the most powerful tools." – Unnamed Defense Analyst

The Shadow of Secrecy

The journey through the glitter conspiracy has taken us down some surprising & unsettling paths. Delving into Chapter 6, we confront one of the most intriguing & quite frankly, alarming aspects of this enigma: the military applications of glitter.

When I first started this investigation, one of the theories driving the conspiracy was that glitter plays a significant role in military technology seemed almost laughable to me. Glitter, after all, is tied to celebrations, crafts, & fashion—not national security.

However, the deeper I dug, the more I realized that this seemingly benign material has far-reaching implications.

The history of military innovation is full of unexpected developments. Glitter, with its unique reflective properties, fits snuggly into this category. It turns out that the same qualities making glitter so appealing in consumer products also make it valuable in military applications.

Deep Roots

What really surprised me was how deep glitter's roots are entwined in modern military technology. It is used not only for simple camouflage but also in sophisticated ways like anti-radar measures & signal disruption.

The idea that a material so closely linked to fun could have such a serious side was a revelation. It made me wonder what else we might have overlooked that could have hidden, more sinister uses.

As I continued to investigate, I uncovered evidence that glitter is used in a variety of defense-related technologies. For instance, it is employed in stealth technology to scatter radar signals, making it harder for enemy systems to detect aircraft.

This alone highlights glitter's strategic importance in military settings. In addition its reflective properties are used in signal jamming and electronic warfare to disrupt communications & confuse enemy sensors.

"The secret of freedom lies in educating people, whereas the secret of tyranny is in keeping them ignorant." –Robespierre

Trivial or Significant

One of the most intriguing aspects of this investigation was discovering how glitter is used in the field of missile defense. By mixing glitter into chaff (a countermeasure used to mislead radar-guided missiles), military forces can protect valuable assets from attacks.

The notion that such a simple material could play a crucial role in safeguarding lives and equipment is both fascinating and unsettling. It underscores the dual nature of technology—how something designed for one purpose can be repurposed for another, often with profound implications.

The more I learned about these applications, the more I realized that my initial impression about this micro precision plastic material was way off the mark –glitter should in no way be considered as 'trivial!' In fact it appears to be more important than anyone seems to want to acknowledge and recognize.

"The surest way to suppress truth is to claim that it is already known." – Anonymous

Classified Uses

It most likely has been 're-classified' as a 'dual-use' technology—one that can be employed for both civilian and military purposes.

This is a haunting theme that recurs more than the public is aware of throughout the entire history of technology initially developed by 'civilians.'. Glitter has been co-opted by the military-industrial complex in ways that are both ingenious and frightening. This realization led me to question what other hidden uses might this seemingly innocuous material might have?

As we identify and examine the various ways that glitter is utilized by the military, you have to keep an open mind & think about the bigger picture. This chapter will explore how glitter is used in a variety of defense technologies, from stealth planes to electronic warfare.

We'll highlight the scientific principles that makes glitter so effective in these applications as we contemplate the ethical and strategic considerations that arise from its use.

"If you want to keep a secret, you must also hide it from yourself." – George Orwell

Questions Worth Asking

As we strive to uncover the truth, we must remain aware of the widespread secrecy and deliberate misinformation that are hallmarks of many government and military operations. These military applications, In the context of the glitter conspiracy, raise some serious questions.

What is the true extent of glitter's use in military applications and why has it been kept hidden from the public? What are the national security implications of this secrecy and how could that have an influence on shaping public awareness?

The answers aren't easy. Sure, on the one hand, there are legitimate reasons for keeping certain aspects of military technology confidential. National security depends on maintaining a strategic advantage, and revealing too much information about specific technologies could compromise that advantage.

But on the other hand, the public has a right to know how taxpayer dollars are being spent and what risks are being taken on their behalf.

"The power to question is the basis of all human progress." – Indira Gandhi

A Compelling Argument

As we delve into the military applications of glitter, we uncover another significant thread in the tapestry of the glitter conspiracy.

The extent to which glitter is utilized in defense applications reveals a startling depth to this seemingly benign material's influence.

From camouflage techniques to advanced communication systems, glitter's role in military technology is both intricate and pervasive. This realization lends a new level of validity and credibility to the theory that glitter's true purpose extends far beyond its surface appearance.

The more we uncover, the more compelling the argument becomes that glitter is not just a decorative element but a crucial component in modern warfare and defense strategies.

This insight transforms our understanding, making the glitter conspiracy not only plausible but profoundly significant in the context of national and global security.

"The more you know, the more you realize you don't know." — Aristotle

Method Mirror

As we transition from our exploration of the glitter conspiracy's broader implications, it's time to dissect some of the specific applications that glitter might have within the military and defense sectors. My investigation required a systematic approach, where each potential use was thoroughly examined, questioned, and then analyzed.

To guide you through this complicated process, I've chosen a structured format that mirrors my investigative process. Each section will begin with a Conspiracy Question—a prompt designed to challenge conventional thinking and explore the possible applications of glitter in military operations.

This question sets the stage for an in-depth summary analysis, by first defining the concept and then reducing it down into key points that illustrate how glitter could be so utilized, and for final consideration I present the evidence I found that supports or refutes these claims.

"A mirror can reflect only what it sees; the truth it shows depends on where you stand." – Unknown

An Informed Personal Opinion

After an objective presentation of this evidence, I offer a balanced Answer which is based on my personal opinion and arrived at after making a determination as to feasibility and the practical limitations of these ideas and the potential resulting consequences.

By framing each military application within this format, we will not only identify possibilities but also be able to determine the 'basis of reality' that stands in support of these theories.

This methodical approach will ensure that we remain critical in our analysis as we continue to sift through speculation in our effort to uncover the truth. It will enable us to view glitter not in the conventional context of a decorative substance but rather as a potential tool in the arsenal of a modern defense strategy—one that could be far more important than you might have ever imagined!

With that in mind, let's dive into the specifics, beginning with the intriguing possibility of the use of glitter in Rocket Science!

"Answers lead to questions, and questions to doubt." — Philip K. Dick

Enhanced Rocket Fuel Additive

Question: *Could glitter be the key to making rockets more efficient and powerful?*

Using glitter in rocket fuel is quite a topic of debate. The idea is that glitter's reflective properties could boost combustion efficiency, making for a more steady burn. This means rockets could go farther, faster, and hit targets better.

Combustion Efficiency: Mixing glitter with rocket fuel might enhance the burn rate. The shiny surface helps spread heat evenly, leading to better combustion. So rockets not only become more powerful but also reliable.

Temperature Regulation: The reflective quality of glitter could keep rocket engines cool, avoiding overheating and failures. By bouncing heat off critical parts, glitter could extend engine life & cut down on malfunctions.

Increased Thrust: Better combustion means more thrust. Rockets could go higher and faster—key for military use where speed & precision matter a lot.

Evidence Against: Some argue that while the theory sounds good, it may be tough to pull off practically.

Foreign particles in rocket fuel could clog lines or harm engines. Plus, lots of testing would be needed to ensure safety.

Answer: *While the idea of using glitter as a rocket fuel additive is intriguing, it remains largely speculative. More research and testing are needed to determine its viability in real-world applications.*

Battlefield Communication Disruption

Question: *Could the release of glitter-like materials be used to jam enemy communications?*

The concept of using glitter to disrupt electronic signals on the battlefield is both fascinating and feasible.

The reflective properties of glitter could interfere with radio waves, GPS signals, and other forms of electronic communication, rendering enemy systems inoperative.

Signal Reflection: Glitter particles reflect and scatter communication signals like an electronic "chaff."

Blocking Signals: This can jam radio communications & GPS signals, stopping enemies from coordinating or targeting specific places.

Temporary Deployment: Glitter's quick & temporary use is a plus. Drones or aircraft can spread it over a wide area fast, creating instant disruption.

Versatility: Glitter can mess up other electronic gear too. For instance, creating a "cloud" that blocks enemy drone signals makes them easy targets.

Evidence Against: Critics say its effectiveness depends on many factors—particle size, how it's spread & weather conditions being most important. .

Additionally, there are concerns about the potential environmental impact of dispersing large quantities of glitter over a battlefield.

Answer: *Using glitter for battlefield communications signal jamming is possible but needs more R&D to understand its full potential and practical limitations. There are also concerns about the potential environmental impact of dispersing large quantities of glitter over a battlefield.*

Visibility Enhancement for Targeting

Conspiracy Question: *Could a glitter-based cloud make enemy drones and missiles easier to target and destroy?*

In modern warfare, the ability to detect and neutralize incoming threats is crucial. Glitter could be used to create a reflective cloud that could enhance the visibility of enemy drones, missiles, and other projectiles, making them easier to track and destroy.

Reflective Coating: When drones or missiles pass through a glitter cloud, they get a shiny coat making them visible to radar and detection systems for better targeting.

Enhanced Tracking: The extra visibility helps tracking systems be accurate in hitting incoming threats—cutting down damage risks.

Strategic Deployment: Spreading glitter clouds where you expect enemy action could create a tactical advantage. By creating zones of enhanced visibility, military forces could gain a significant tactical edge, deterring enemy attacks while improving the effectiveness of their countermeasures.

Evidence Against: Some argue that the practicality of deploying glitter clouds in a combat environment is questionable.

Factors such as wind, weather, and terrain could affect the dispersion and effectiveness of the glitter cloud . Additionally, the long-term environmental impact of such a tactic remains a concern.

Answer: *The use of glitter to enhance the visibility of enemy drones and missiles is an innovative concept with potential tactical benefits. However, its practical application would require careful consideration of environmental and logistical challenges. Countermeasure Development*

Countermeasure Development

Conspiracy Question: *Could glitter be used in advanced countermeasures against electronic warfare?*

Glitter's reflective qualities might help make countermeasures protecting military gear from electronic attacks work better by disrupting enemy signals near critical infrastructure.

Electronic Shielding: Glitter could create shields blocking/reflecting signals—keeping sensitive info safe from interference.

Deception Tactics: Glitter could also fool enemy sensors with false signals/decoys hiding real assets from attacks.

Versatile Deployment: With its use flexibility (aerosol sprays or drones), it quickly adapts to evolving threats as countermeasure options increase.

Evidence Against: Although promising, challenges remain: deployment precision/environmental impacts are major hurdles needing careful evaluation before full implementation.

Answer: *Glitter-based countermeasures need extensive testing but represent an innovative approach that needs refining first for reliability across varied scenarios.*

"Sometimes the questions are complicated and the answers are simple, but often, it's the other way around." — Dr. Seuss

Eyes Wide Open

From enhancing rocket fuel to disrupting enemy communications, glitter's applications in military and defense contexts reveals a material with remarkable versatility & strategic value that plays a transformative role in modern warfare.

However, the practical implementation of these concepts requires further research and development, as well as careful consideration of ethical, environmental, and logistical factors.

Reflecting on the many-sided uses of glitter in military & defense sectors has been nothing short of eye-opening.

At first, it seemed almost bizarre to think something as seemingly frivolous as glitter could have serious applications in warfare. But, as we've uncovered throughout this chapter, the reality is far more complex.

The depth of innovation and the strategic applications of glitter reveal just how far technology can evolve from its original, simple uses.

"Technology is a useful servant but a dangerous master." – *Christian Lous Lange*

Reflective Innovations

To me, one of the most surprising aspects of my investigation was discovering the extent to which glitter's reflective properties have been harnessed for military camouflage and signal disruption. It's fascinating to think that the same material used in arts and crafts can also play a crucial role in keeping soldiers hidden and communications secure. The ingenuity behind these applications highlights a pattern we've seen throughout history: military necessity driving technological innovation.

You have to admit that the use of glitter in advanced camouflage techniques is particularly compelling. The idea that the reflective particles of glitter can help break up and disrupt the outlines of soldiers and equipment in various environments is both simple and brilliant.

It reminded me of how sometimes the simplest solutions can have the most significant impacts. It's a testament to human ingenuity that something so small can make such a big difference in the field of stealth and concealment.

"Sometimes, the simplest solutions are the most profound." — Unknown

The Cutting Edge

What I found equally intriguing during the course of my investigation into military and defense applications was glitter's role in disrupting electronic signals. In an age where warfare is increasingly dependent on technology, the ability to interfere with enemy communications can provide a significant tactical advantage.

Learning about how glitter is used to scatter radio waves and obscure communication channels made me take pause and serious consider how unaware people are about the myriad of unseen ways technology directly influences modern warfare.

The revelation that glitter is used in high-tech weaponry like laser-guided munitions was also a bit of a game-changer. It's incredible to consider that these tiny particles can enhance the accuracy of some of the most advanced weapons in existence.

This not only widened my view of what glitter can do, but also showed that there is no limitation to how far military researchers are willing to go to improve the precision & effectiveness of weapons used in combat.

Tying It All Together

Understanding the military applications of glitter adds another thread we need to untangle in order to reveal the image hidden within the intricate tapestry of the glitter conspiracy. It underscores the versatility and potential impact of this material, and demonstrates how something as small as a micro-precise piece of glitter could have such a giant impact on a nation's warfare & defense strategy.

This knowledge may be crucial to providing context and insight into why this seemingly innocuous substance has been shielded in such secrecy.

*"**At the end of the day, the questions we ask determine the answers we find.**"* –Adams

One Step Closer

In our quest to uncover the truth behind the glitter conspiracy, this chapter adds yet another missing piece to the puzzle. It demonstrates that glitter's significance is not limited to commercial and decorative uses and its tentacles extend well into the realms of defense and national security.

This understanding brings us one step closer to 'seeing the big picture' and allows us to grasp the full scope and magnitude of the 'major players' involved in the glitter conspiracy.

"The greatest discoveries come from tying together what you already know." – Unknown

Chapter Question

Could glitter be the secret weapon that turns the tide in modern warfare?

Answer

Yes! Glitter's unique properties have the potential to revolutionize military tactics and technologies. From improving the efficiency of rocket fuel to creating electronic countermeasures and enhancing the visibility of incoming threats, glitter could play a crucial role in a variety of future combat scenarios.

Final Thought to Think About

"Every piece of the puzzle that you find is one less piece you need to discover." – **Unknown**

Chapter 7

Environmental & Health Crisis

'Wake up, sheeple!'

Chapter Question

Could the glitter we cherish today be the silent killer of tomorrow, threatening the health of every living being on the planet?

In recent years, a disturbing trend has come to light—micro-plastics, those tiny fragments of plastic less than five millimeters in size, have infiltrated nearly every corner of our planet. They are in our soil, waterways, oceans, and even the air we breathe. What's even more alarming is that glitter is a significant contributor to this widespread pollution.

"The environment is where we all meet; where we all have a mutual interest; it is the one thing all of us share." – Lady Bird Johnson

The Ubiquity of Micro-plastics

Imagine this: a product we use for decoration and fun has turned into a global pollutant that knows no bounds. The small size and lightweight nature of glitter make it particularly insidious, allowing it to travel vast distances and contaminate ecosystems far and wide.

In one shocking study, micro-plastics were detected in 83% of tap water samples around the world, underscoring just how pervasive this problem has become.

What's truly unsettling is that these micro-plastics are not confined to urban or industrial areas. They have been found in some of the most remote and pristine environments on Earth, from the icy reaches of the Arctic to the deepest ocean trenches. It's a stark reminder that no ecosystem is safe from this growing menace.

The relentless production and consumption of plastic products, including glitter, have led to a situation where micro-plastics are now entering the food chain and potentially affecting every living organism on the planet.

"We are living on this planet as if we had another one to go to." – Terri Swearingen

Environmental Impact

Glitter, with its dazzling sparkle, is far more dangerous than it appears. Made from a combination of plastic and reflective materials like aluminum, it doesn't break down easily. Instead, it persists in the environment, adding to the already staggering levels of micro-plastic pollution. As these particles accumulate, they pose a serious threat to aquatic life, soil health, and even air quality. The environmental impact of glitter is profound, and it's only just beginning to be understood.

--Aquatic Life

When glitter and other micro-plastics find their way into our waterways, they become a silent killer for marine life. Fish and other creatures often mistake these tiny particles for food, leading to physical blockages in their digestive systems. But the harm doesn't stop there—micro-plastics can also carry toxins, which latch onto the particles and are ingested along with them. The consequences can be fatal.

Research has shown that micro-plastics can also affect the reproductive systems of fish, leading to reduced fertility and declining populations. It throws off the delicate balance of the food chain and poses a grave threat to biodiversity.

--Soil Health

Micro-plastic pollutants in soil affect its structure & fertility. These particles can mess with water retention and hinder plant growth by blocking nutrient absorption. Studies have shown they can change the physical and chemical properties of soil—potentially leading to reduced crop yields & compromised soil health.

Long-term consequences are still being studied but the impact of micro-plastic pollutants could be significant for agriculture.

--Air Quality

It's not just land and sea that are affected—micro-plastics have been found in the very air we breathe. In urban areas, where plastic use is high, these particles can become airborne and contribute directly to air pollution. Breathing in micro-plastics can pose a serious health risk, particularly for those people with a pre-existing respiratory condition like asthma.

The full extent of the health implications of airborne micro-plastics is still unknown, but the presence of these particles in the air might actually end up eventually leaving us all 'breathless'.

Human Health Concerns

When I first delved into the world of micro-plastics, I was shocked to discover just how pervasive these tiny particles have become—not just in our environment but within our bodies.

The thought of glitter—a material so often associated with fun and celebration—becoming a potential health hazard is both unsettling and an eye-opener.

Studies have shown that micro-plastics are finding their way into some of our most vital organs, including our lungs, liver, and kidneys. The idea that these particles, which are often invisible to the naked eye, could be causing inflammation and oxidative stress within our bodies is deeply concerning.

The implications here are profound—what we breathe in, what we eat, and what we drink may be silently undermining our health. We're only now beginning to grasp the full scope of the damage micro-plastics can cause.

"Our planet's alarm is going off, and it is time to wake up and take action!" —
Leonardo DiCaprio

Inside All of Us

To think that these particles are being discovered even in human stool samples is a stark reminder of how pervasive micro-plastic pollution has become. It's not just out there in the oceans or floating through the air—it's in us.

Recent research has even raised concerns about how micro-plastics could disrupt hormone production, leading to imbalances and other related health issues. This isn't just about a minor inconvenience; we're talking about potential disruptions to fundamental biological processes. The evidence is starting to pile up: these tiny particles can act as carriers for harmful chemicals, including persistent organic pollutants (POPs), which can accumulate in the body over time, leading to chronic health conditions.

As we continue to explore the long-term effects of micro-plastic exposure, my evaluation of the initial data and preliminary findings indicate that we very well could be on the brink of a global public health crisis.

"You can ignore reality, but you cannot ignore the consequences of ignoring reality." – Ayn Rand

Government Secrecy & Cover-Ups

As I dug deeper into the implications of micro-plastic pollution, another familiar disturbing trend emerged—the role of government secrecy. Despite mounting evidence pointing to the dangers of micro-plastics, it seems that some governments are more interested in downplaying these risks than addressing them head-on. Access to vital information is often restricted, and research into the health effects of micro-plastics is grossly underfunded.

Whistleblowers have come forward with stories of data manipulation and the suppression of scientific research that highlights the risks associated with micro-plastics. This lack of transparency isn't just about keeping people in the dark; it's about actively undermining public trust and making it nearly impossible for concerned citizens to take meaningful action.

One can't help but wonder if this secrecy is designed to protect industrial interests and prevent public panic.

"Secrecy, being an instrument of conspiracy, ought never to be the system of a regular government." – Jeremy Bentham

The Forever Product

Glitter, despite its charm and allure, harbors a dark side—it's what I've come to call a "forever product." Unlike organic materials that break down and return to the earth, glitter is composed of plastic and metal, materials that defy nature's recycling processes. Imagine a product so resilient that it could potentially outlast the civilization that created it!

This durability, while seemingly impressive, is actually a big part of the problem. Another part is Glitter's metal coating which enhances its resistance, allowing it to withstand even the harshest environmental conditions.

Once these tiny particles are released into the environment, they're not going anywhere. They persist, contributing to the endless cycle of pollution that plagues our planet.

It's a sobering thought that something as small and seemingly harmless as glitter could have such a lasting impact. We must vocalize the urgent need for a sustainable alternative and insist on better waste management practices and more effective recycling programs.

"Forever is an illusion. What matters is now." – Eleanor Brown

The Invisible Threat

If you think glitter's environmental impact is bad, brace yourself for the next revelation: micro-plastic pollutants are literally everywhere. From bottled water to table salt, and even in the very air we breathe, these tiny particles have infiltrated and infected every aspect of our lives.

The scariest part? We've all been exposed to these pollutants, whether we realize it or not. The long-term effects of this long-term exposure are still 'being determined', but initial findings indicate that 'worse case scenarios' cannot be ruled out.

Studies have shown that micro-plastics can act like a "toxic taxi," picking up and transporting harmful chemicals wherever they go. This makes them even more dangerous, potentially turning them into carriers of disease capable of 'total' environmental destruction.

As I dug deeper into this issue, I couldn't help but think of the quote: "In nature, nothing exists alone." This statement rings truer than ever as we realize how interconnected our health is with the health of our environment.

"No amount of evidence will ever persuade an idiot." – Mark Twain

Exaggerated Health Risks?

Let's not mince words—the potential health risk posed by micro-plastics is absolutely terrifying. Just imagine these particles accumulating in your bloodstream, leading to chronic inflammation, increased cancer risks, and hormonal disruptions that could affect everything from fertility to fetal development.

It's like something out of a science fiction nightmare, except it's all too real.

As if there isn't enough proof of the potential danger to our long term health there's growing evidence that micro-plastics can cross biological barriers, such as the blood-brain barrier, raising serious concerns about their impact on neurological health.

The idea that these tiny invaders could be contributing to cellular damage and genetic mutations is nothing short of alarming. It's clear that we need a new approach to dealing with this crisis, and we need it NOW!

"We cannot solve our problems with the same thinking we used when we created them." – Albert Einstein

Information Control

As I delved deeper into the micro-plastic crisis, I couldn't help but notice a disturbing pattern: governments worldwide aren't just downplaying the risks associated with micro-plastic pollutants—they're actively hiding the truth. We live in a time when information is more accessible than ever, yet critical data about the dangers of micro-plastics is being withheld, classified, or simply left unfunded. It's a chilling reminder of how power operates behind closed doors.

This suppression isn't just about keeping the public calm; it's also about protecting industrial interests. The fact that we've seen whistleblowers silenced and research stifled only adds to the growing body of evidence that there is a deliberate effort being made to keep us in the dark.

We also cannot ignore the widespread reporting that details government censorship and the public intimidation and official 'responses' to the concerns of scientists and environmental advocates. These are individuals who have dared to speak out about the dangers associated with micro-plastic pollutants, only to have their voices silenced.

The lack of regulatory action and the snail's pace of policy development to address micro-plastic pollution are not just bureaucratic inefficiencies—they are deliberate and coordinated efforts to ensure that the uninformed public remains just that: uninformed.

"To keep people unaware is to control them. Knowledge is power, and those who hold it rule the world." – Unknown

A Call to Action

Given the severe implications of micro-plastic pollution, it is imperative that we take action now. We must reduce the use of glitter and other micro-plastics, support research into their health effects, and demand transparency from governments and industries immediately.

The public must be informed and mobilized in order to address this silent killer. Community-led initiatives and grassroots movements could play a vital role in raising awareness and advocating for policy changes.

"The only way forward, if we are going to improve the quality of the environment, is to get everybody involved." – Richard Rogers

Still Puzzled

In this chapter, we've delved deep into the environmental and health crises stemming from micro-plastic pollution, with glitter being a significant contributor to this pervasive issue. We explored how micro-plastics infiltrate nearly every part of our ecosystem—from the air we breathe to the water we drink, and even the food we consume.

The findings are alarming, pointing to a potential public health crisis that is being systematically downplayed and obscured by both government and industry entities. This suppression of vital information is not just a side issue; it's a central piece of the glitter conspiracy puzzle.

Coming Into Focus

As we look ahead to subsequent chapters, we will continue to peel back the layers of this conspiracy, examining how these micro-plastic pollutants fit into the larger narrative of secrecy and manipulation. We will explore the role of how the glitter conspiracy ties into broader themes of environmental degradation, public health risks, and the erosion of public trust.

*"**When you look directly into the light, you are blinded. But when you look indirectly, the truth is illuminated.**" —Proverb*

Chapter Question

Could the glitter we cherish today be the silent killer of tomorrow, threatening the health of every living being on the planet?

Answer

Yes, the glitter we cherish today could indeed be a silent killer of tomorrow, posing significant health risks and environmental damage. The evidence suggests that micro-plastics, including glitter, are pervasive pollutants that threaten the health and wellbeing of every living thing on the planet.

Final Thought To Think About

"We do not inherit the earth from our ancestors, we borrow it from our children." – **Native American Proverb**

Chapter 8

Historic Threads of Deception

'Big Brother is watching you'

Chapter Question

Could the government and powerful industries be deliberately hiding the truth about glitter for sinister purposes?

As we dig deeper into the enigmatic world of the glitter conspiracy, it becomes crucial to pause and reflect on the broader tapestry of history, where conspiracies—once dismissed as mere fantasies—have been validated by the cold, hard truth. This chapter, therefore, is not just an exploration of past conspiracies but a necessary groundwork for understanding the true scope and implications of the glitter conspiracy.

"The best way to keep a prisoner from escaping is to make sure he never knows he's in prison" - Fyodor Dostoevsky

Lessons Learned

Why should we delve into the history of conspiracies? The answer is simple yet profound: history is our greatest teacher. By examining the past we will uncover patterns, motivations, and methods that have been used repeatedly by those in power to manipulate, deceive, and control.

These historical lessons are not just academic exercises rooted in the past; they are the keys to unlocking the mysteries of the present. The glitter conspiracy, with its web of secrecy and intrigue, should not be viewed as an isolated incident but rather as a continuation of a long-standing tradition of covert operations, hidden agendas, and suppressed truths.

A History of Hidden Agendas

In this chapter, we will take a comprehensive look at some of the most significant conspiracies throughout history—those that were initially scoffed at, ridiculed, or dismissed, only to be later validated by undeniable evidence. From government cover-ups and corporate malfeasance to clandestine operations and assassinations, these conspiracies have shaped the world we live in today.

By understanding how these events unfolded, we can better appreciate the potential reality behind the glitter conspiracy and its far-reaching implications.

"The farther backward you can look, the farther forward you are likely to see." –
Winston Churchill

Tricks of The Trade

Each conspiracy we examine will be presented not just as a historical event but as a critical piece of the puzzle, offering insights into the mechanisms of secrecy and control. We will explore how these conspiracies were uncovered, the resistance faced by those who sought the truth, and the eventual revelations that shocked the world.

Through this exploration, we will draw parallels to the glitter conspiracy, highlighting how similar tactics of obfuscation and misdirection are being employed today

Consider Watergate —a conspiracy that, at first glance, seemed too far-fetched to be true. Yet, the persistence of investigative journalists exposed a web of corruption and deceit that reached the highest levels of government.

Meaningful Moments

As we delve into these historical conspiracies, I will also share my personal experiences as an investigative journalist—moments of revelation, frustration, and triumph that have shaped my understanding of the lengths to which those in power will go to protect their secrets.

My journey has not been an easy one, filled with dead ends, threats, and the constant battle against the forces of suppression. But it is these very challenges that have fueled my determination to uncover the truth, no matter how deeply it is buried.

"Little by little, one travels far." –
J.R.R. Tolkien

Connect The Dots

The information in this chapter will be presented in a way that educates & engages you, the reader in a critical examination of the evidence. Each conspiracy will be dissected, analyzed, and connected back to the central theme of the glitter conspiracy, showing how history is not just a series of disconnected events but a continuous thread that weaves through time, linking past and present in a complex web of cause and effect.

By the end of this chapter, you'll understand how conspiracies are not just products of paranoid minds but are often rooted in real events that have been deliberately distorted and intentionally hidden from the public.

You will also see how the glitter conspiracy ties into this broader historical context, serving as a stark reminder that the truth, no matter how well concealed, has a way of coming to light

"You can't connect the dots looking forward; you can only connect them looking backward." – Steve Jobs

.A Trip Worth Taking

So, as we embark on this remember that the pursuit of truth is a journey worth taking, no matter how treacherous the path may be. The lessons we learn from the past will be our guide as we continue to unravel the glitter conspiracy, shedding light on the darkness that has enveloped this mysterious and potentially world-altering phenomenon.

Operation Northwoods

Operation Northwoods was a plan cooked up by the U.S Department of Defense in1962. It was a false flag operation. The idea was to do bad stuff in America and blame on Cuba. They wanted to make people mad they could go after Fidel Castro's government. They thought about hijacking planes, sinking boats with Cuban refugees, and starting violent attacks in American cities.

Years later, in the 1990s, Operation Northwoods' documents got declassified and people were shocked! They showed how far the government was prepared to go to achieve its goals. People got upset and wondered how the government could go to such extremes in their effort to sway public opinion in order to gain support to start a war with Cuba?

In the context of the glitter conspiracy, Operation Northwoods serves as a stark reminder that governments are capable of conceiving and nearly executing elaborate deceptions to manipulate public opinion. Just as Northwoods aimed to manufacture consent for war, the glitter conspiracy could involve similar tactics to obscure its true purpose and influence global agendas.

The Pentagon Papers

The Pentagon Papers refers to a secret report about the U.S.'s political and military role in Vietnam from 1945 to 1967. In 1971, Daniel Ellsberg leaked them to the New York Times and boy, did it cause a stir! The papers showed that the U.S. government lied to both Congress and the public about what was really going on in Vietnam. They were ramping up the war while telling everyone they wanted peace.

This leak led to a huge Supreme Court case. The court said that the press had the right to publish these papers even if the government didn't like it one bit. Public trust in government took a nosedive & anti-war feelings grew even stronger. The Pentagon papers made clear that officials knew the war couldn't be won but kept sending soldiers anyway!

The exposure of this conspiracy had a big impact on public opinion leading to widespread protests and a shift in public opinion about the Vietnam War. This case shows how important the role of the media is in holding the government accountable and underscores the importance of a free press in a democracy.

The Cambridge Analytica Scandal

In 2018, Christopher Wylie blew the whistle on Cambridge Analytica, a data analytics firm, who had harvested the personal data of millions of Facebook users without their knowledge or consent.

The data was used to influence voter behavior in political campaigns, including the 2016 U.S. presidential election and the Brexit referendum. The scandal exposed the dark side of data privacy and the extent to which personal information could be manipulated for political gain..

The fallout from the scandal was immense, with Facebook CEO Mark Zuckerberg testifying before Congress and facing intense questioning about the company's data policies.

The scandal also prompted widespread public awareness about the importance of data privacy and the risks of sharing personal information online. In the context of the glitter conspiracy, the Cambridge Analytica scandal highlights the potential for data and information to be covertly weaponized.

The Boeing 737 MAX Scandal

Boeing's 737 MAX scandal is pretty recent but very serious! After two crashes in 2018 and 2019 investigations revealed that Boeing had deliberately concealed critical information about the new flight control system, MCAS, from regulators and pilots.

The company prioritized cost savings and a faster certification process over public safety, leading to the loss of 346 lives. Boeing's internal communications showed that some employees were aware of illegal safety violations and inferior substitute material issues but said and did nothing and repeatedly failed to contact or inform the appropriate regulatory authorities.

The Federal Aviation Administration (FAA) was also criticized for its oversight failures and reliance on Boeing to conduct it's 'independent' safety assessments.

The scandal highlighted significant flaws in the relationship between regulators and the industries they oversee, where corporate secrecy policies and the resulting conflicts of interest could lead to dangerous outcomes.

In the context of the glitter conspiracy, the Boeing 737 MAX scandal underscores the critical importance of transparency and regulatory oversight. Just as Boeing's concealment of safety issues led to tragic outcomes, the hidden uses and undisclosed impacts of glitter could have similarly devastating consequences if left unchecked.

The ExxonMobil Climate Change Controversy

Back in 2015 news hit hard: ExxonMobil knew fossil fuels caused climate change since '70s! Internal docs revealed company scientists who had researched global warming extensively acknowledged the risks and dangers associated with burning fossil fuels, yet publicly downplayed or funded campaigns casting doubt on consensus findings around its impact on climate change.

The revelation of ExxonMobil's actions sparked outrage and led to several lawsuits accusing the company of misleading the public and investors about the environmental and financial risks associated with fossil fuels.

The controversy highlighted the role of corporate secrecy in shaping public perception and policy on critical issues like climate change.

ExxonMobil's actions have had long-lasting implications for the global climate debate, contributing directly to delays in instituting any meaningful policy action and by consequence exacerbating the impacts of climate change.

The company's efforts to suppress critical information and intentionally mislead the public underscore the potential dangers of corporate secrecy and the need for greater transparency in the energy industry.

In the context of the glitter conspiracy, the ExxonMobil climate change controversy illustrates the profound dangers of corporate secrecy in shaping public perception and policy.

Just as ExxonMobil downplayed the risks of climate change, the true implications of the glitter conspiracy may be similarly obscured, with potentially dire consequences for the environment and society.

"A nation of sheep will beget a government of wolves." – Edward R. Murrow

The NSA's PRISM Program

The PRISM Program operated by the National Security Agency (NSA) was one of the most significant revelations made by a system consultant Edward Snowden back in 2013. This top-secret program was designed to collect vast amounts of data from major tech companies like Google, Apple, and Facebook, including emails, videos, and voice chats.

The program's existence was kept hidden from the public and even from many government officials, raising concerns about privacy and government overreach.

Critics argue that PRISM infringed on civil liberties, with the government essentially spying on its own citizens under the guise of national security.

The secrecy surrounding the program also meant that there was little oversight or accountability. This lack of transparency has fueled mistrust in government agencies and their respect for personal privacy. Despite the uproar caused by the revelation, the government defended PRISM, claiming it was 'a national security issue' and essential for counterterrorism efforts.

However, the balance between security and privacy remains a contentious issue, with many advocating for stricter regulations and oversight.

In the context of the glitter conspiracy, the NSA's PRISM program serves as a stark reminder of how far governments can go in the name of national security, often operating in the shadows with little public oversight. The secretive nature of the glitter conspiracy could involve similar tactics, with information being withheld under the guise of protecting national interests.

The Flint Water Crisis

The Flint water crisis is an example of government and industrial negligence and secrecy causing widespread harm. In 2014, the city of Flint, Michigan, switched its water supply to the Flint River to cut costs.

However, the water was not properly treated, leading to lead contamination that affected thousands of residents, including many school-aged children.

For months, officials denied and downplayed the severity of the contamination despite mounting evidence and public outcry. Whistleblowers, independent researchers, and investigative journalists eventually brought the crisis to light.

In revealing the extent of the cover-up and the total failure of every local, state and federal agencies and organizations to take any action to protect the health of the public even after they knew it was necessary and by virtue of their job titles and descriptions were 'legally required' to do so.

In the context of the glitter conspiracy, the Flint Water Crisis highlights the devastating consequences of government and industrial secrecy. Just as the failure to act in Flint led to widespread harm, the concealment of critical information about the glitter conspiracy could result in severe, unforeseen public health and environmental impacts.

"The great masses of the people will more easily fall victims to a big lie than to a small one." – Adolf Hitler

A Balancing Act

The cases presented in this chapter paint a vivid picture of how secrecy can shape and, at times, distort reality. Governments and industries, each wielding significant power, often use secrecy as a tool to control narratives, protect interests, and, ostensibly, safeguard the public.

The increased level of secrecy in both government and industry contexts reveals a troubling pattern. One that history has revealed to us many times that we would be fools to ignore.

Whether it's the suppression of scientific data, the concealment of harmful product information, or the clandestine operations of intelligence agencies, these practices can erode public trust and compromise ethical standards.

The balancing act between necessary confidentiality and public transparency is a challenge that requires one to maintain an equilibrium, and leaning too far towards secrecy can 'knock us off-balance' and lead to abuses of power and a total disconnect between the 'governed' and 'the 'government.'

"Balance is not something you find, it's something you create." – Jana Kingsford

"

Future Reflections

As we reflect on these examples, it becomes clear that vigilance and accountability are paramount. Whistleblowers, investigative journalists, and an informed public play crucial roles in challenging the status quo and bringing hidden truths out from the shadows and into the light.

The drive for secrecy in both government and industry highlights the ongoing struggle between power and accountability. While some level of secrecy may always be necessary, it is the duty of an informed and vigilant populace to ensure that such secrecy is justified and intended to serves the greater good rather than the narrow interests of any individual or entity.

By learning from the past, we can build a better, brighter future where transparency and truth are not the exception but the norm.

We live in a world where secrecy is paramount to those in power, and they are ruthless in protecting it." – Peter F. Hamilton

Conclusion

As we have seen, the history of government and industry secrecy is long and troubling. From the Manhattan Project to the NSA surveillance programs, there are countless examples of the lengths to which those in power will go to keep the truth hidden. The glitter conspiracy fits neatly into this pattern, raising serious questions about what the government and this powerful industry might actually be trying to hide and conceal from the public.

This chapter has highlighted the historical context of government and industry secrecy, provided case studies of past conspiracies, and presented stories and personal anecdotes that should persuade even the most skeptical reader to acknowledge that 'something is not as it seems.

Preface for Chapter 9 Transition

In the upcoming chapter the questions we have explored throughout this book will finally find their answers. The enigma surrounding the mysterious buyers, the unexplained secrecy, and the potential high-stakes applications of glitter all point to a surprising revelation.

"The most dangerous untruths are truths slightly distorted."

— George Christopher Lichtenberg

Chapter Question

Could the government and powerful industries be deliberately hiding the truth about glitter for sinister purposes?

Answer

Yes, it's plausible that the government and powerful industries are hiding the truth about glitter for sinister purposes. The history of deceit and control by those in power supports this theory. But what if there's a deeper layer to this conspiracy?

Could there be a noble reason behind the secrecy? The truth, as you'll discover in the next chapter, is more complex and unexpected than it seems.

Final Thought To Think About

"Sometimes the truth is so precious, it must be hidden in plain sight."

— **George Orwell**

Chapter 9

Unveiling The Conspiracy Mastermind

"It's all connected."

Chapter Question

Could the mysterious buyer of glitter be using it to advance a secretive high-tech project that has the potential to change the world?

In this chapter, you will learn how visionary thinking has transformed industries and explore the broader implications for the future of technology and humanity. Most importantly, we will finally reveal the name of the person behind the glitter conspiracy, shedding light on the mastermind who has been orchestrating it all.

The greatest threat to our planet is the belief that someone else will save it." – Robert Swan

Rebel With A Cause

As we've wandered through the twists and turns of the glitter conspiracy maze, one name keeps popping up, a name that really needs no introduction. But I must tread lightly here. The clues have been subtle, most of the time they were hidden in plain sight, and only through careful observation and relentless pursuit of the truth was I able to piece together the astonishing connection.

It is a connection that may reshape our understanding of the world and its future once it is fully understood. You have traveled with me through the dark corridors of history, where conspiracies once thought to be mere fantasy have eventually been revealed to actually be regarded as an undeniable fact. Now, we stand on the brink of uncovering one the most extraordinary revelations of all time.

At the heart of this discovery is a man known for his relentless drive, his visionary ideas, and his unyielding belief in the potential of technology to advance humanity. But beneath the surface of his well-documented achievements lies a secret—one that has been meticulously guarded and cleverly disguised.

MVP

It is a secret that, if true, not only links him to the glitter conspiracy but positions him as a key player in a larger, more complex tapestry of global influence and control. He is a figure who has never been content with following the rules; instead, he has always sought to rewrite them.

Digging into this person wasn't easy. The connection seemed too wild at first. Honestly, I thought it was just a crazy theory. I originally dismissed the idea as nothing more than a wild theory, the kind of conjecture that borders on the absurd. However, as I delved deeper, gathered evidence and corroborated facts, I began to see an undeniable pattern emerge.

"There are no secrets that time does not reveal" –Jean Racine

Not So Hidden Agenda

This revelation rocked me to my core. I had long admired this individual for his bold vision and his contributions to advancing technology. His companies have been at the forefront of some of the most significant technological advancements of our time. But this new information suggests something far more complex and far-reaching.

It suggested a hidden agenda, one that could only be fully understood by retracing his steps back through time that led him to where he is today. It also hinted at a major power shift quietly unfolding behind closed doors.

As I continued my investigation, I realized that his rise to power was not as straightforward as it seemed. The lucrative government contracts that fueled his early success were a double-edged sword. On one hand, they provided the necessary funding and resources to bring his ambitious projects to life. On the other hand, they tied him to the very entities he would later seek to outmaneuver.

Power Play

But as his power grew, so too did his ability to act alone and operate independently. Today, he has more capabilities in certain areas then they do and no longer has to rely on them or ask for any favors, instead they rely on him and ask him to do favors for them.

This shift in power dynamics is by no means an accident. It is the result of years of careful planning, strategic alliances, and, perhaps, a little bit of 'the shiny, sparkly stuff.'.

The idea that something that appears to be so simple could play such an essential role in this grand scheme may seem laughable at first. But when you consider the innovative ways in which this micro-material can be utilized—from enhancing solar panels to improving the efficiency of advanced optics—the pieces begin to fall into place

It's been a wild ride connecting these dots but worth every minute as we edge closer to revealing a truth that might reshape our worldviews forever!

"To acquire power, one must possess it." – Maurice Maeterlinck

The Real Cost of Progress

My personal ethics and beliefs have always driven me to seek out the truth, to uncover the hidden narratives that shape our world. I believe in the power of technology to advance mankind and lead us into a future that is defined by innovation and progress. Yet, as I dig deeper into this story, I'm starting to wonder about the 'real' cost of all this progress. What's the actual price we pay for advancements that come shrouded in such secrecy?? And who really benefits from these hidden agendas?

One thing's for sure: the glitter conspiracy is more than just a mystery waiting to be solved. It is a window into the soul of a man who has redefined what it means to be a visionary and a leader. His journey from obscurity to becoming a global influence is a testament to his ingenuity, but it is also a reminder that power, once attained, can be wielded in ways that are not always visible to the naked eye.

As I prepare to reveal the identity of this individual, I am acutely aware that we are not just uncovering a conspiracy—we are unraveling the very fabric of power in the 21st century.

"Power is always dangerous. Power attracts the worst and corrupts the best." —
Edward Abbey

The Grand Master

As we pull back the layers of this tangled web, it's becomes clear that the enigmatic figure we've been circling around is not just any innovator; he's the Grand Master of Innovation. His influence? HUGE. His reach? Seemingly without end. And his vision for humanity's future? It's nothing short of revolutionary.

At first I didn't want to believe it, the idea seemed too far-fetched, too grandiose, even for someone like him. But the more evidence I unearthed, the more undeniable it became.

Every piece of the puzzle just seemed to fit, and when you assembled them all together his name appeared right smack in the middle of the big picture.

Still, I needed more than circumstantial evidence. I needed confirmation—something solid, something irrefutable. And so, I reached out directly to 'first person' sources, those who had worked closely enough with him and who have witnessed the lengths he would go to in order to achieve his vision.

"What you get by achieving your goals is not as important as what you become by achieving your goals." – Zig Ziglar

Verification & Validation

One source, someone close to him who had been personally involved in some of his most secretive projects, finally broke the silence and reluctantly confirmed that the individual I had identified was in fact 'the mystery' man behind the glitter conspiracy.

In response to my questions as to why it has been so 'hush – hush' and shrouded in such secrecy, he replied "He doesn't want to have to ask anyone for permission," his voice tinged with both admiration and unease.

"He doesn't want to be told that he isn't 'allowed' to try and do this", he further explained, and only after he proves it can be done and that it will work will he tell the world that he has solved the global climate crisis ."

"The question isn't who is going to let me; it's who is going to stop me." – Ayn Rand

A Control Freak

I struggled at first with this idea—it clashed with his public image as a visionary savior, all about advancing tech for the greater good. But the more I learned, the clearer it got: his public face was just part of the story.

Underneath lies a much more complicated guy—an individual whose ambitions stretch beyond any single project or company. He isn't just keen on innovation; he craves control—control so thorough that even top governments are in check.

This realization was confirmed by a second, equally credible source within the government, "The contracts, the funding—they were just stepping stones he used to build what he needed, but now he's the one holding all the cards." he explained. "He has everything he needs to do it and we don't so we've agreed to 'assume ignorance' so he can give it a try.".".

Right smack in the middle of this wild tale, at the heart of this glitter conspiracy, there's a guy who's made it his life's work to push limits. The clues that I dug up—rocket ships, global satellite networks, quiet power plays—all point to one man whose name is synonymous with the cutting edge of technology. Who is this mystery man I am talking about? It's none other than Elon Musk.

"We are all in the gutter, but some of us are looking at the stars." – Oscar Wilde

The Glitter King

As we explore the origins of his empire, we will uncover the connections that link Musk's public successes to secretive, glitter-fueled projects that gave him an unparalleled competitive advantage..

And, most importantly, we will ask the questions that few dare to pose: What is Elon Musk really up to? And where will his ambitions take us next? What are the implications of a man with such power, & influence, who operates in the shadows without any oversight?

"He who controls others may be powerful, but he who has mastered himself is mightier still." – Lao Tzu

SolarCity: Harnessing the Power of the Sun

Founded in 2006, SolarCity marked Musk's initial foray into renewable energy, driven by a bold vision: to make affordable solar energy accessible to all. The company set out to provide cost-effective solar solutions to both homeowners and businesses, pushing the boundaries of what was possible in the industry.

While it may seem improbable, the use of glitter in solar panels played a pivotal role in this mission. The reflective properties of glitter, when integrated into the surface of photovoltaic cells, significantly enhanced light absorption, leading to increased energy production.

This innovation not only made SolarCity's products more efficient but also more appealing to consumers, contributing to the company's rapid success. The merger of SolarCity with Tesla in 2016 further advanced Musk's vision of a sustainable future.

The incorporation of glitter-enhanced solar panels into Tesla's energy solutions has been instrumental in the development of the Tesla Solar Roof, a product that seamlessly blends aesthetics with high functionality.

Tesla: Revolutionizing the Auto Industry

Tesla, founded in 2003, has transformed the automotive industry with the introduction of electric vehicles (EVs) that rival traditional gasoline-powered cars in both performance and sustainability.

One of the significant challenges Tesla faced was improving battery life and efficiency. By integrating glitter into the battery designs, its unique properties allowed for a more uniform distribution of energy, reducing the risk of overheating and extending the batteries' lifespan.

This breakthrough was a game-changer, enabling Tesla to achieve impressive milestones such as the Model S's 300-mile travel range and the Roadster's rapid acceleration. These innovations positioned Tesla at the forefront of the EV market, solidifying Musk's reputation as a pioneering force in sustainable transportation.

"Discovery consists of seeing what everybody has seen and thinking what nobody has thought." – Albert Szent-Györgyi

SpaceX: Reaching for the Stars

Founded in 2002, SpaceX has since established itself as THE global leader in space exploration and technology. Musk's ambitious goal for SpaceX is to make space travel more accessible and affordable, with the ultimate aim of enabling human colonization of Mars.

Surprisingly, glitter has played a role in this mission, particularly in the development of rocket fuel additives that ensure a more stable and efficient burn rate. This advancement has allowed SpaceX rockets to achieve higher altitudes and more precise trajectories, reducing launch costs and increasing success rates.

Moreover, the reflective properties of glitter have contributed to the development of advanced thermal protection systems for re-entry vehicles. By incorporating glitter into heat shields, SpaceX has enhanced their ability to reflect and dissipate heat, ensuring the safe return of spacecraft.

"Invention is the mother of necessity." Veblen

Neuralink: Connecting Minds

Neuralink, founded by Musk in 2016, is focused on the development of cutting-edge brain-machine interface technology. While glitter might seem an unlikely component in neuroscience, its applications in micro-engineering and material science have potential implications for Neuralink's next-generation optical interface micro-components.

Glitter's conductive properties can enhance the performance of neural interfaces, allowing for more precise signal transmission and reducing the risk of interference. This innovation could pave the way for breakthroughs in transforming the future of human-machine integration.

"First we build the machines, then they build us." –Marshall McLuhan

The Boring Company: Digging Deep

Founded in 2016, The Boring Company aims to revolutionize urban transportation by creating underground systems designed to alleviate traffic congestion. Glitter's potential applications in tunneling equipment include enhancing the durability and efficiency of cutting tools.

By incorporating glitter into the design of cutting heads, The Boring Company could reduce wear and tear, increasing the lifespan and efficiency of their equipment—critical factors in the success of large-scale tunneling projects.

"The deeper you dig, the more treasures you'll uncover." – Matshona Dhliwayo

Starlink: Connecting the Globe

Starlink, a project under SpaceX, is designed to provide global internet coverage through a network of low Earth orbit satellites. Glitter's reflective properties are utilized in the design of these satellite components to improve signal transmission and reduce interference.

By enhancing the efficiency of Starlink satellites, Musk is bringing his vision of global connectivity closer to reality, promising to bridge the digital divide and provide internet access to even the most remote corners of the world

"The true sign of intelligence is not knowledge but imagination." – Albert Einstei

A True Visionary

As we delve deeper into the innovations and groundbreaking achievements driven by Elon Musk, it's essential to understand the underlying motivations that have fueled this relentless pursuit of progress.

The technologies and companies we've explored are not just the products of a brilliant mind—they are the manifestations of a broader vision that seeks to redefine the future of humanity.

To fully grasp the implications of Musk's work, we must first explore the mindset that drives him, the principles that guide his decisions, and the dreams that keep him pushing forward, even in the face of seemingly insurmountable challenges.

Understanding this mindset will set the stage for what comes next: the unveiling of "Mission X," a concept that ties together Musk's ambitions and the mysterious threads we've been following throughout this journey.

"Understanding others is knowledge; understanding yourself is enlightenment." –
Lao Tzu

The Musk Mindset

Elon Musk's drive to save humanity and advance civilization can be found at the core of all his ventures.. His commitment to sustainability, innovation, and exploration is clearly evident and led to the development of renewable energy solutions, made the mass production of electric vehicles possible, and the commercialization of space travel a reality.

His belief in technology and its potential to solve the world's 'problems' has been a guiding principle throughout his career

Musk's mindset can be characterized as a relentless pursuit of progress combined with an innovative spirit and a willingness to take risks. His dedication to propelling humanity towards becoming an interplanetary species reflects his broader vision of ensuring the survival and advancement of human civilization.

His ability to think outside the box and push the boundaries of what is possible has made Musk a unique figure in the world of technology and innovation.

"The future belongs to those who believe in the beauty of their dreams." – Eleanor Roosevelt

Mission X: The Ultimate Goal

The culmination of Musk's efforts is Mission X, a secretive project aimed at reducing global warming and solving the climate crisis. By using glitter to create a reflective barrier in the Earth's atmosphere, Musk hopes to reduce the planet's temperature and mitigate the effects of climate change.

This ambitious plan involves deployment of large volumes of glitter into space using SpaceX rockets and utilizing the Starlink satellite network to contain, maintain, and monitor the reflective shield.

The glitter particles, designed to reflect away the heat generating frequencies of sunlight, would help cool the Earth's surface and slow the progression of global warming.

Musk's unique qualifications, capabilities and resources including his satellite network, rocket technology, and financial resources, make him the ideal candidate to spearhead this mission.

"Some are born great, some achieve greatness, and some have greatness thrust upon them." – William Shakespeare

The Concept

Mission X refers to a plan by Elon Musk where he intends to deploy a layer of holographic glitter just above Earth's atmosphere to reflect sunlight and mitigate global warming. This glitter will act as a solar shield, bouncing back harmful rays and helping to cool the planet.

This concept is grounded in proven scientific principles and could be achieved using existing materials and available technologies.

Selective Reflective Material

Holographic glitter can be engineered to allow certain frequencies of light to 'pass through' while blocking or 'reflecting back' other specific frequencies of light. '

Heat' is primarily contained in the red and infra-red part of the spectrum. A holographic glitter product designed for this project would allow the other 'frequencies' and colors in the spectrum of light to pass through down to the surface of the planet but would reflect the red and infra-red part of the spectrum away from the earth back towards the sun and out into outer space.

Holographic Glitter Parameters

1. .Material Composition:.

Polyester Film:. The base of holographic glitter is often a thin polyester film, known for its durability and resistance to environmental degradation.

Reflective Coating:. A layer of aluminum or other reflective metals enhances the film's ability to reflect light. This coating can be adjusted so that it reacts differently when it interacts with different wavelengths of light.

2. .Optical Engineering:.

Interference Patterns:. Holographic glitter utilizes interference patterns created by layering materials with different refractive indices. These patterns reflect certain wavelengths of light while allowing others to pass through.

Frequency Selection:. The glitter is fine-tuned to reflect infrared and other heat-related wavelengths back into space while allowing beneficial light frequencies to pass through. This selective reflection helps reduce the heat trapped by Earth's atmosphere.

Deployment and Maintenance

Deploying glitter into the stratosphere involves several challenges. SpaceX, with its advanced rocket technology, is uniquely positioned to address these challenges.

1. .Rocket Technology:.

Reusable Rockets:. SpaceX's Falcon 9 and Starship rockets, designed for reusability, make frequent launches economically feasible. These rockets can carry large payloads of glitter to the desired altitude.

Precision Deployment:. The ability to deploy payloads with high precision ensures that the glitter is released in precise amounts to form an even solar shield

2. .Electromagnetic Fields:.

Satellite Network:. Starlink satellites can generate a low-intensity electromagnetic field to keep the glitter particles suspended and evenly distributed just above the atmosphere. This field acts like a net, preventing the glitter from clumping together or drifting out of position.

Real-Time Monitoring:. The satellites provide real-time monitoring and adjustments, ensuring the glitter remains effective over time.

Durability and Longevity

One of the biggest questions is how long the glitter can remain effective. Factors such as degradation from UV radiation, atmospheric conditions, and physical wear must be considered.

1. .UV Resistance:.

Protective Coatings:. The glitter can be coated with UV-resistant materials to prolong its lifespan. These coatings prevent the breakdown of the reflective properties and maintain the integrity of the glitter.

Simulation Studies:. Simulations and laboratory tests suggest that with proper coatings, the glitter can remain effective for several years before needing replenishment.

2. .Environmental Impact:.

Biodegradability:. Research into biodegradable materials for the glitter is ongoing. Using biodegradable polymers could mitigate any potential environmental impact, ensuring that the glitter breaks down harmlessly after its effective lifespan.

.Feasibility and Challenges

While the concept is scientifically plausible, several challenges must be addressed:

1. .Volume and Coverage:.

Quantity Needed:. The sheer volume of glitter required to cover a significant portion of the atmosphere is immense. Calculations must be made to determine the optimal density and distribution.

Uniform Coverage:. Achieving uniform coverage is crucial for the effectiveness of the solar shield. This requires precise deployment and continuous adjustments.

2. .Cost and Funding:.

Economic Viability:. The cost of manufacturing and deploying the glitter, along with the maintenance of the satellite network, must be weighed against the potential benefits.

Funding from governments, private investors, and international coalitions could make the project feasible.

3. .Potential Risks:.

Weather Patterns:. The introduction of a reflective layer just above the atmosphere could potentially alter weather patterns. Detailed climate models must be developed to predict and mitigate any adverse effects.

Space Debris:. The presence of glitter in the upper atmosphere could contribute to space debris. Measures must be taken to ensure that the particles do not interfere with satellites or other space missions.

.Seasonal Deployment.

Deploying the glitter during hurricane season could have significant benefits. By lowering global ocean temperatures by 3-5 degrees, the conditions required for hurricane formation could be prevented.

1. .Temperature Regulation:.

Ocean Cooling:. The reflective properties of the glitter can reduce the amount of heat absorbed by the oceans. Lowering ocean temperatures can disrupt the formation of hurricanes, which rely on warm water to gain strength.

Simulation Models:. Climate simulations indicate that even a small reduction in ocean temperatures can significantly impact hurricane formation and intensity.

2. .Global Impact:.

Weather Stability:. The deployment of glitter can contribute to global weather stability. This can prevent the loss of life and property associated with severe weather events.

Should You Chose To Accept It

Mission X presents a visionary approach to combating climate change. By leveraging advanced materials, rocket technology, and satellite networks, Elon Musk's initiative could offer a viable solution to one of humanity's most pressing challenges.

However, careful planning, extensive research, and international collaboration are essential to address the scientific, logistical, and ethical questions that arise.

While Musk has privately acknowledged his involvement in Mission X, he continues to publicly deny any knowledge or participation. This strategic denial allows him to avoid regulatory hurdles and opposition from governments and environmental groups. By staying in the shadows, Musk can continue to advance Mission X without drawing unnecessary attention, allowing the project to progress unimpeded until it is ready to reveal its full potential to the world.

"The power of one man or one woman doing the right thing for the right reason, at the right time, is the greatest influence in our society." – Jack Kemp

Conclusion and Reflection

After 15 months of exhaustive research I have confidently concluded that Elon Musk is the mysterious mastermind behind the glitter conspiracy. His utilization of glitter in various high-tech applications over the years gave him a competitive edge and helped him advanced his objectives.

From solar panels to electric cars, to rocket fuel additives and optical brain-machine interfaces, glitter has been a critical component in ensuring Musk's success. The versatility and unique properties of glitter have allowed Musk to innovate in ways previously thought impossible.

Musk's dedication to saving humanity and ensuring the survival of civilization is evident in his ambitious projects and innovative solutions. Mission X represents the pinnacle of his efforts, a bold plan to combat climate change and end global warming.

Musk is uniquely qualified to undertake such a mission. His vision, resources, and technological capabilities position him as the one person who could actually 'save the planet.'

"The true sign of intelligence is not knowledge but imagination." – Albert Einstein

Chapter Question

Could the mysterious buyer of glitter be using it to advance a secretive high-tech project that has the potential to change the world?

Answer

Yes, the mysterious buyer of glitter is using it to advance a secretive high-tech project that has the potential to change the world. Elon Musk's innovative use of glitter in various applications throughout his career has given him a competitive edge and enabled him to pursue ambitious projects aimed at saving humanity and addressing global challenges. T his is just another chapter in Musk's ongoing mission to push the boundaries of what's possible in order to secure a sustainable future for generations to come

Final Thought To Think About

"The people who are crazy enough to think they can change the world are the ones who do." –
Steve Jobs

Chapter 10

Public Reaction & Government Response

"They control the media"

Chapter Question

Could public awareness of a clandestine project designed to combat climate change lead to unforeseen global repercussions?

Public reactions to high-stakes conspiracies often unfold in stages, starting with skepticism and evolving into widespread debate and sometimes panic. When a significant figure like Elon Musk is rumored to be behind a secret project, these reactions can become even more intense. This chapter explores the anticipated public reactions and the varying responses from government and industry leaders .

"The public's reaction to a sudden truth is often disbelief followed by indignation." —
Unknown

Progression & Protocol

As we close the chapter on Elon Musk's groundbreaking ventures and the revelations surrounding them, we enter a critical phase of this journey: the public's reckoning with the truth. The disclosure of a clandestine project like Mission X, with its far-reaching implications for climate change and global power dynamics, is certain to stir the 'world-wide' collective consciousness in profound ways.

Public reaction to such a revelation is rarely linear or predictable, yet it often follows a recognizable pattern—a standard protocol that has been repeated time and time again throughout history.

In this chapter, we will examine the progression and various degrees of the public's anticipated reaction and response to the disclosure of Mission X. From initial skepticism to eventual acceptance or outright panic, these stages reflect the psychological and social dynamics that come into play when the public is confronted with a truth that has been hidden in plain sight.

"The public is wonderfully tolerant. It forgives everything except genius." –O. Wilde

More Likely Than Not

The responses identified are 'highly probable' and are based on a long history of the public's reaction to similar high-stakes disclosures. Whether it's the release of the Pentagon Papers, the fallout from the Cambridge Analytica scandal, or the unveiling of government surveillance programs, these events illustrate the historic precedent that the public's journey from disbelief to action is both inevitable and predictable.

In this chapter, I have dissected each stage of the anticipated reaction, and speculate as to how it will unfold in the public sphere and what it might reveal about our society.

I also explore how governments and industry leaders are 'most likely' to respond, as they seek to strike a delicate balance between managing the perception of the public and confronting the very real consequences of such a revelation. A proper protocol that is initiated after this type of disclosure is atypically designed to contain panic, control the narrative, and mitigate the damages.

"It is not the truth that matters, but the truth one perceives." – Henry Kissinger

Fasten Your Seatbelt

The uniqueness of Mission X and its potential impact on the future of humanity may very well lead to a global response and reaction that is unlike anything we have ever seen before.

Sit back, strap in and brace for impact as we venture into 'unknown territory' and navigate the complex interplay between truth and perception, where the outcome is uncertain and the stakes are 'as high as the sky.'

"The world is not dangerous because of those who do harm but because of those who look at it without doing anything." – Albert Einstein

Stages of Public Reaction

1. Initial Skepticism

When the rumors about Mission X first surface, the public is likely to respond with a mix of disbelief and intrigue. Conspiracy theories often face initial skepticism, especially when they involve high-profile individuals like Elon Musk. However, given Musk's reputation for groundbreaking projects, some will find the theory plausible.

2. Growing Curiosity

As more details emerge the interest and curiosity of the public will surge. People will start asking questions, sharing info on social media, and offer opinions as to success or failure of the project as they hunt for any evidence they can find that will 'support' their position.

Influencers and journalists will dig deep for details as they conduct 'online and in-person' investigations, dissecting and examining every aspect of the project. .

Involvement will increases interest, which in turn will increase awareness, resulting in an increase in news coverage that grows and becomes a 'media magnet' that attracts the attention of the public.

3. Public Outcry

Once credible sources start confirming aspects of the conspiracy, the public outcry will grow. Environmentalists, tech enthusiasts, and conspiracy theorists will all have their say.

Some will praise Musk for his willingness to accept the challenge, while others will express concern over the secrecy and potential risks involved in Mission X.

4. Demand for Transparency

Demand for transparency will mount, with the public pressuring Musk, government officials, and industry leaders to explain why such an important project was shrouded in secrecy and what its true impacts could be.

This mounting pressure will result in a series of official statements and an onslaught of press conferences, leading to a seemingly perpetual cycle of explanations and clarifications.

"Public opinion is no more than this: what people think that other people think." – Alfred Austin

Stages of Government Response

1. Initial Denial

Government and industry leaders will initially categorically deny any knowledge of Mission X, and dismiss it as a 'baseless conspiracy theory' in an attempt to downplay the significance of this 'erroneous information.' This is a common stall tactic used to buy the time needed to formulate and initiate an effective media 'counter-strategy' that seeks to degrade the situation's current status and alter its projected trajectory.

2. Acknowledgment with Limited Information

As evidence mounts, the government will be forced to acknowledge the existence of a project but will provide limited information.

They might cite national security concerns as a reason for the secrecy, trying to placate the public while still keeping the majority of the details under wraps.

3. Controlled Disclosure

When undeniable proof is presented, the government will be forced to shift to a controlled disclosure strategy.

They will release carefully curetted information to the public, emphasizing and exaggerating the positive aspects of the project while minimizing potential risks.

This stage aims to regain public trust and control the narrative.

4. Full Disclosure and Justification

Eventually, under immense public pressure, full disclosure will be made.

The government and Musk will justify the secrecy by highlighting the potential benefits of Mission X, such as combating climate change and reversing global warming.

They will argue that the project's success depended on maintaining confidentiality in order to prevent 'outside' interference and ensure its viability.

"People do not believe lies because they have to, but because they want to." Muggeridge

Musk's Personal Response

1. Categorical Denial

Elon Musk will initially deny any knowledge of Mission X. Known for his candid communication style, Musk will use his social media platforms to dismiss the rumors of his involvement with this project as 'another unfounded conspiracy theory.'

2. Reluctant Admission

As more tangible, 'undisputable' evidence emerges, Musk will reluctantly admit to the existence of the project. He will explain his reluctance to disclose Mission X was due to his innate fear of failure. The potential of a 'negative' response and unfavorable reaction to the announcement of the Mission X project could have generated widespread global resistance and induced public panic.

3. Highlighting Benefits

Musk will emphasize the innovative nature of Mission X, explaining how it could significantly mitigate climate change. He will share some of the technical details, showcasing the project's feasibility and potential for success, while downplaying or even refusing to mention any 'unforseen' consequences that could arise.

4. Call for Support

Musk will call for public support, urging people to understand the necessity of secrecy for such a groundbreaking project. He will appeal to their sense of global responsibility, positioning Mission X as a crucial step toward saving the planet.

> *"What you do makes a difference, and you have to decide what kind of difference you want to make."* – Jane Goodall

Negative Potential Consequences

1. Investor Backlash

Investors in Tesla and other Musk-led ventures might become disenchanted, fearing that Mission X is too great a responsibility for one individual.

This could lead to a decrease in stock prices and loss of investor confidence.

2. Legal Challenges

Disaffected third parties, including international competitors and environmental groups, could seek legal action to halt Mission X.

They might request a cease and desist order from the World Court, citing the potential environmental risks and the lack of any regulatory oversight.

3. Political Opposition

-The United Nations and various world governments might oppose Mission X, fearing that it could disrupt global power dynamics.

They might argue that such a significant world-wide project should not be under the control of a single individual, no matter how well-intentioned.

Additionally, concerns over sovereignty and the potential for one entity to wield too much influence over global climate strategies could further fuel political resistance.

4. Public Panic

The revelation of Mission X could lead to public panic, particularly if the project's goals and methods are not clearly communicated.

Fear of the unknown and the potential of a negative outcome could cause widespread anxiety and opposition.

Misinformation and sensationalized media coverage could exacerbate these fears, leading to protests and demands for immediate cessation of the project.

5. Scientific Skepticism

Some scientists might estimate the project's success rate at only around 16%, arguing that it is too risky to pursue.

They could highlight the fact that the 'risk's are very 'real' and the benefits are only 'imagined', and when taken into account the risk's outweigh the benefits. However the 'bottom line could bottom out' when you factor in 'costs' associated with unintended environmental consequences or failure to achieve the desired impact.

"The reaction to the truth is usually dependent on how it was delivered." –
Unknown

Positive Potential Consequences

1. Global Collaboration

The disclosure of Mission X could lead to a surge in global collaboration. Scientists, engineers, and environmental activists might rally around Musk, offering their expertise and resources to ensure the project's success.

2. Increased Investment

Awareness of Mission X could attract new investors who are passionate about combating climate change. This influx of funding could accelerate the project's development and implementation.

3. Technological Advancements

Mission X could spur technological advancements in related fields, such as renewable energy, satellite technology, and environmental monitoring. These innovations could have far-reaching benefits beyond the scope of the project itself.

4. Public Mobilization

Environmental activists and concerned citizens might mobilize to support Mission X.

Grassroots campaigns, social media movements, and public demonstrations could create a groundswell of support, pressuring governments and industry leaders to back the project.

"What we need is more people who specialize in the impossible." – Theodore Roethke

An Awakening

Upon navigating our way successfully through the complex and multifaceted reactions to the disclosure of Mission X we are left with the feeling that this is far from over.

The stages of public reaction—from skepticism to curiosity, fear, and ultimately, understanding—are just the beginning of a broader societal transformation.

While the initial response may be fraught with uncertainty and conflict, the long-term implications of such a revelation offer a glimpse into a brighter future where technology and transparency drive positive change. Once fully understood, the positive potential consequences of Mission X might be a global awakening that acknowledges the power of innovation to solve the most pressing challenges of our time.

This project, born from secrecy and shrouded in controversy, has the capacity to inspire a new era of collaboration and progress, where the boundaries of what is possible are continually expanded.

The ripple effects of this realization could foster a renewed commitment to sustainability, encourage greater public engagement with scientific endeavors, and even reshape the relationship between the people 'in' control and the people 'under' control.

"Just as ripples spread out when a single pebble is dropped into water, the actions of individuals can have far-reaching effects." – Dalai Lama

Adaptability

The transition from shock to acceptance, from fear to hope, will require not only time but also a collective willingness to embrace change. The true potential of Mission X will only be realized if humanity can move beyond the initial turmoil and towards a future where such innovations are no longer denied and dismissed and instead are acknowledged, accepted and supported.

As we turn the page to the next chapter, we leave behind the initial responses & reactions and enter a period of reflection & adaptation. The undeniable implication of Mission X is its potential to shape the course of global events and directly influence the trajectory of human progress.

Hopefully what comes next is a story of resilience, innovation, and the enduring power of the human spirit to turn even the most disruptive truth into an opportunity for growth and advancement.

"It is not the strongest of the species that survive, nor the most intelligent, but the one most responsive to change." – Charles Darwin

Chapter Question:

Could public awareness of a clandestine project designed to combat climate change lead to unforeseen global repercussions?

Answer:

Yes, public awareness of Mission X could lead to significant global repercussions, ranging from heightened scrutiny and political opposition to widespread support and mobilization for climate action.

The secrecy surrounding Mission X was essential to prevent premature interference, ensuring the project's success while minimizing potential panic and resistance.

By examining the past, we learn that the government and industry leaders often engage in secrecy for the greater good, even if it means navigating a minefield of public distrust and controversy.

Final Thought To Think About

"The best way to predict the future is to create it." –* **Peter Drucker**

Chapter 11

A Call to Action

"Wake up, sheeple!"

Chapter Question

Can we, as a global community, mobilize effectively to support a clandestine mission aimed at combating climate change?

In the face of global challenges, the power of collective action cannot be overstated. This chapter aims to provide a roadmap for mobilizing support for Mission X, leveraging grassroots activism, scientific collaboration, & corporate engagement. My personal goal is to do whatever I can to ensure that Elon Musk's vision for addressing climate change is not only realized but is also shielded from potential obstacles posed by governments or third parties.

What you do today can improve all your tomorrows." – Ralph Marston

Taking A Stand

As we arrive at this final chapter, we stand at the culmination of a journey that has taken us through the intricate web of secrets, power struggles, and groundbreaking innovations that have come to define the glitter conspiracy.

From the initial skepticism and disbelief to the growing realization of the truth, we have uncovered a narrative that challenges our understanding of both technology and humanity's future. Now, with all the pieces of the puzzle laid before us, the question remains: what do we do with this knowledge?

"Never underestimate the power of a small group of committed people to change the world. In fact, it is the only thing that ever has." — Margaret Mead

Personal Recommendation

In this chapter, I shift the focus from uncovering truths to taking action. It is my opinion that the time has come to consider how we, as a global community, can mobilize a world-wide movement in support of Mission X.

This chapter will present my personal recommendations for how we can work together to ensure that Musk's visionary project is not only supported but is also protected from those who might seek to derail it. From grassroots activism to corporate collaboration, these strategies aim to harness the collective power of individuals and institutions to bring about meaningful change.

The Choice Is Yours

I recognize that not everyone may share my enthusiasm for this cause. Some may view the secrecy surrounding Mission X with suspicion, others may have ethical concerns, and still others may believe that such a project is too ambitious to succeed. These objections are valid, and I respect the diversity of opinions that this complex issue inevitably generates.

However, I firmly believe that the potential benefits of Mission X far outweigh the risks, and that this project represents a critical opportunity to address the environmental challenges that threaten our world.

"Simple solutions seldom are. It takes a very unusual mind to undertake analysis of the obvious." — Alfred North Whitehead

Put Our Hands Together

If you, like me, see the promise in Musk's vision and believe in the power of collective action, then I invite you to join me in taking these next steps. The recommendations that follow are designed to empower individuals and communities to make a tangible difference.

Together, we can ensure that this mission—conceived in secrecy but poised to change the world—receives what it needs to succeed!

"Great things are not done by impulse, but by a series of small things brought together."
Vincent Van Gogh

Grassroots Mobilization

1. Local Activism

--Form local activist groups that can raise awareness about Mission X within communities. This can include organizing events, distributing flyers, and engaging with local media.

Partner with community groups, church organizations, and youth clubs to spread the word.

-These entities have established networks that can be instrumental in rallying support.

2. Educational Outreach

Engage with educational institutions, from high schools to universities, to integrate discussions about Mission X into their curriculum. Host workshops and seminars to educate students about the project's significance and the role they can play.

Encourage student groups to form Mission X clubs that can serve as hubs for discussion, planning, and action.

3. Social Media Campaigns

--Leverage social media platforms to create viral campaigns that inform and inspire.

Use hashtags like :#MissionX, #ClimateAction, and #SupportMusk to unify efforts across different platforms.

- Create engaging content such as videos, info graphics, and memes that highlight the urgency of climate action and the potential of Mission X.

"Success is the sum of small efforts, repeated day in and day out." – Robert Collier

Corporate Engagement

1. Branding and Promotion

- Encourage corporations to integrate the Mission X icon onto their branding. This can include adding the logo to their products, websites, and promotional materials.

- Highlight companies that publicly support Mission X, creating a positive feedback loop where consumer support is driven by corporate endorsement.

2. Corporate Social Responsibility (CSR)

- Urge companies to align their CSR efforts with the goals of Mission X.

This can involve funding research, supporting environmental initiatives, or directly contributing to the project.

- Showcase success stories of businesses that have made significant contributions to Mission X, inspiring others to follow suit.

"What is right is not always popular and what is popular is not always right." – Albert Einstein

Scientific Collaboration

1. Expert Coalition

Form a coalition of scientific experts from diverse fields to provide insights, conduct research, and validate the project's feasibility.

Create forums where ideas, suggestions, and recommendations can be discussed and shared openly, ensuring transparency and collaboration.

2. Advisory Board

- Recommend the formation of an advisory board under the United Nations or the World Council, consisting of eminent scientists, environmentalists, and policy makers. This board can monitor the progress of Mission X, provide guidance, and ensure accountability.

- Advocate for the advisory board to have the authority to combat and neutralize any attempts by governments or third parties to obstruct the project.

"Collaboration allows us to know more than we are capable of knowing by ourselves." —
Paul Solarz

Public Engagement

1. Online and In-Person Engagement

- Use real-time language interpreters to enable worldwide formation of support groups. These groups can pledge their support for Mission X and demonstrate their commitment through online and offline activities.

Encourage individuals and organizations to host events, webinars, and discussions to keep the momentum going. Engage with influencers and thought leaders to amplify the message and create a global community of advocates.

2. Promotional Slogans and Appeals

- Develop catchy slogans and appeals to raise awareness and rally support. Examples include:

- **"Glitter the Future: Support Mission X"**

- **"Shine Bright for a Cooler Planet"**

- **"Mission X: Our Last, Best Hope"**

   "Cool the Earth, Light the Way: Mission X"

"Small deeds done are better than great deeds planned." — Peter Marshall

Full Circle

As we bring this journey to a close, it's important to reflect on the path we've traveled together. What began as a series of questions and whispers in the dark has unfolded into a narrative that challenges everything we thought we knew about the world around us.

The glitter conspiracy, once dismissed as just another wild theory, has revealed itself to be a key piece in a much larger puzzle—one that involves some of the most powerful forces on our planet, all converging at this critical moment in the history of mankind.

As we stand at the precipice of this revelation, it's clear that our understanding of power, influence, and secrecy has been irrevocably transformed. What seemed improbable has become undeniable, forcing us to reconsider the very foundations of what we believed to be true. Now, with the full picture before us, we must decide how we will move forward in this newly illuminated world.

"It is not in the stars to hold our destiny but in ourselves." – William Shakespeare

I Salute You

To those who have been labeled as conspiracy theorists, mocked for questioning the status quo, this book stands as a testament to your courage and tenacity. Your willingness to dig deeper, to ask the uncomfortable questions, and to connect the dots has brought us closer to the truth.

In a world where information is often controlled, manipulated, or buried, the role of the truth-seeker is more vital than ever. Your instincts to challenge 'authorities', to ask hard questions, and to never settle for the easy answer have been vindicated.

As we unearth the layers of deception and secrecy, your perseverance has illuminated the dark corners where others feared to tread. You have shown that the pursuit of truth is not just an intellectual exercise but a moral obligation, one that demands resilience, critical thinking, and an unwavering commitment to uncovering what lies beneath the surface. For this, I salute you.

"The most dangerous man to any government is the man who is able to think things out for himself." – H.L. Mencken

Icon Of Innovation

The symbol of Mission X should become a beacon of hope and determination, appearing on websites, products, and clothing, signaling a united front in the fight against climate change.

Now, as we look forward, the power to shape the future lies in our hands. The revelations uncovered here are not just stories to be told— they are calls to action. The time has come to move beyond awareness and into mobilization, to take the knowledge we've gained and use it to create the change we so desperately need.

Mission X, with all its potential, offers a path forward, but it's up to us to ensure that this path is followed with integrity, purpose, and unity.

"Symbols are powerful because they are the visible signs of invisible realities."
– Saint Augustine

In Closing...

As we close this chapter, let us carry forward the lessons learned, the questions raised, and the determination to make a difference. We are living in extraordinary times, where the convergence of technology, power, and secrecy demands our attention and our action.

This book was written with the aim to bring you the information when you needed it most, to shine a light on the shadows where truth often hides. And now, as we stand on the brink of what could be a pivotal moment in human history, remember that this journey is far from over.

Stay sharp. Stay inquisitive. Stay dedicated. In a high-stakes world, the critical information might seem delayed—but **with my help** it always arrives right when you need it—just in time."

"This is not the end. It is not even the beginning of the end. But it is, perhaps, the end of the beginning." – Winston Churchill

Chapter Question

Can we, as a global community, mobilize effectively to support a clandestine mission aimed at combating climate change?

Answer

Yes, by leveraging grassroots activism, scientific collaboration, and corporate engagement, we can effectively mobilize global support for Mission X. This collective effort is essential to overcoming potential obstacles and ensuring the project's success, ultimately safeguarding our planet's future.

Final Thought To Think About

"A small body of determined spirits fired by an unquenchable faith in their mission can alter the course of history."
— **Mahatma Gandhi**

About The Author

Justin Time is a pseudonymous author and investigative journalist whose work delves deep into the labyrinth of the world's most enigmatic and controversial conspiracies. With a career spanning over two decades, Justin has earned a reputation for fearless reporting and an unwavering commitment to uncovering hidden truths.

His investigations have taken him to the shadowy fringes of society, where power and secrecy intersect, and where the stories that shape our world are often concealed from public view.

Background and Experience:

Justin's journey began with a passion for understanding the unseen forces that shape our world. His academic pursuits in secret histories and alternative narratives equipped him with the tools to question official stories and dig deeper into the layers of deception that often surround major global events.

Over the years, Justin has conducted field research in some of the most restricted and clandestine environments, always seeking to bring to light the truths that powerful entities would prefer to keep hidden.

Throughout his career, Justin has published under various pseudonyms, each tailored to the specific investigation at hand. This strategy allows him to approach each subject with a fresh perspective, free from any preconceived biases or judgments that might arise from previous work.

By doing so, he ensures that each investigation stands on its own merits, allowing readers to engage with the material without any preconceived notions.

His work has appeared across a range of platforms, from blogs and social media accounts to underground publications, each contributing to his reputation as a relentless seeker of truth.

Why the Pseudonym?

Operating under a pseudonym isn't just a choice for Justin—it's a necessity. In a world where power often seeks to silence those who challenge it, maintaining anonymity allows him to pursue the truth without fear of retribution.

By staying on the periphery, outside the pull of the center, Justin preserves his objectivity and safeguards his sources. The name "Justin Time" itself is a nod to the urgency and timeliness of his work—bringing critical information to light just when it's needed most.

Justin's use of different names for different investigations also serves to protect the integrity of his work. By not being tied to a single identity, he avoids the pitfalls of becoming a "branded" voice, which can often lead to expectations or biases from both readers and critics.

Instead, he remains an enigma, allowing the focus to remain squarely on the investigations and the truths they uncover, rather than on the person behind the words.

Notable Works and Recognitions:

While Justin Time is the name associated with this particular investigation, his work spans a variety of issues and conspiracies, each tackled with the same rigor and dedication to truth. His investigations have been recognized with numerous awards, such as The Phantom Award for Investigative Journalism.

He has also received recognition and was awarded 'The Cloak and Dagger Prize' for his deep dives into secret government projects and corporate corruption.

Though his previous works may have been published under different names, they all share the same commitment to uncovering the unseen and exposing it to the light.

A Life Dedicated to the Truth:

Very little is known about Justin's personal life, a deliberate choice given the sensitive and often dangerous nature of his work. He is believed to live off-the-grid, dedicating his time to researching and writing about the next big conspiracy that threatens to remain hidden.

His commitment to truth and transparency, coupled with his refusal to be swayed by external pressures, makes him a unique voice in the world of investigative journalism.

. In a world where information is often hidden, he is dedicated to uncovering the unseen and presenting it when you need it --Justin Time!

Legal Disclaimer

Due to the nature of this book and the high likelihood that powerful entities or individuals—especially those with significant financial resources—may attempt to suppress the information presented herein, I am hereby making the following legally binding declaration:

*"The content of *All That Glitters: The Truth Behind the Glitter Conspiracy Revealed* represents the personal opinions, interpretations, and investigative findings of the author, Justin Time.*

This book is intended solely for informational, educational, and entertainment purposes. While every effort has been made to ensure the accuracy of the information presented, the author and publisher disclaim any responsibility for errors or omissions.

The scenarios, characters, and events described in this book are based on research, investigative journalism, and the author's interpretations of publicly available information, anecdotal evidence, and speculative reasoning.

The author does not claim to have direct evidence of any illegal or unethical activities by any individuals or entities mentioned in this book, and the reader is encouraged to conduct their own research and come to their own conclusions.

This book does not constitute legal, financial, or professional advice.

The opinions expressed are those of the author alone and do not reflect the opinions of any organizations, institutions, or individuals with whom the author may be affiliated.

Any resemblance to real persons, living or dead, or actual events is purely coincidental.

The author and publisher assume no liability for any actions taken by individuals or entities based on the information provided in this book.

The author and publisher specifically disclaim any and all liability for any claims or damages that may result from the use or interpretation of the material in this book, including but not limited to legal actions taken by individuals or corporations mentioned within.

Furthermore, the author acknowledges that some information presented in this book may challenge the views of powerful individuals or corporations.

This book is not intended to defame, libel, or slander any person, corporation, or entity, and any assertions made within are based on the author's freedom of expression and the right to investigate and question publicly available information.

By reading this book, the reader agrees that any disputes or legal claims arising from the content within will be resolved through confidential arbitration in a jurisdiction chosen by the author and publisher, with no further recourse to public litigation."